Fab

Also by Neil Gershenfeld

When Things Start to Think

The Nature of Mathematical Modeling

The Physics of Information Technology

Fab

The Coming Revolution on
Your Desktop—from Personal
Computers to Personal Fabrication

Neil Gershenfeld

BASIC
BOOKS

A MEMBER OF THE PERSEUS BOOKS GROUP
NEW YORK

Published by Basic Books,
A Member of the Perseus Books Group

Basic Books are available at special discounts for bulk purchases in the United States
by corporations, institutions, and other organizations. For more information, please
contact the Special Markets Department at the Perseus Books Group, 11 Cambridge
Center, Cambridge MA 02142; or call (617) 252-5298 or (800) 255-1514;
or e-mail special.markets@perseusbooks.com.

Designed by Lovedog Studio

LIBRARY OF CONGRESS CATALOGING-IN-PUBLICATION DATA
Gershenfeld, Neil A.
  Fab : the coming revolution on your desktop—from personal computers to
personal fabrication / by Neil Gershenfeld.—1st ed.
      p. cm.
  Includes index.
  ISBN 0-465-02745-8 (hc. : alk. paper)
  1. Technological forecasting. 2. Computers and civilization. I. Title.

T174.G476 2005
600—dc22

                                            2005000988

              05 06 07 08 / 10 9 8 7 6 5 4

For Kalbag

and

Kelly, Meejin, Dalia, Irene, Saul, Alan, Grace,
Eli, Mel, Kyei, Anil, Frank, Larry, Etienne, Seymour,
David, Kenny, Amy, Haakon, Vicente, Terry,
Sugata, Arjun

and

Lass, Susan, John, Bakhtiar, Amon, Joe, Aisha,
Chris, Caroline, Manu, Sanjay, Debu,
Jørgen, Benn, Milton, Joe, Ike, Scott, Seth,
Alex, Mitch, Marvin

Contents

How to Make . . .

Mainframe computers were expensive machines with limited markets, used by skilled operators working in specialized rooms to perform repetitive industrial operations. We can laugh in retrospect at the small size of the early sales forecasts for computers; when the packaging of computation made it accessible to ordinary people in the form of personal computers, the result was a unprecedented outpouring of new ways to work and play.

However, the machines that make computers (and most everything else) remain expensive tools with limited markets, used by skilled operators working in specialized rooms to perform repetitive industrial operations. Like the earlier transition from mainframes to PCs, the capabilities of machine tools will become accessible to ordinary people in the form of personal fabricators (PFs). This time around, though, the implications are likely to be even greater because what's being personalized is our physical world of atoms rather than the computer's digital world of bits.

A PF is a machine that makes machines; it's like a printer that can print *things* rather than images. By personal fabrication, I mean not only the creation of three-dimensional structures but also the integration of logic, sensing, actuation, and display—everything that's needed to make a complete functioning system. With a PF, instead of shopping for and ordering a product, you could download or develop its description, supplying the fabricator with designs and raw materials.

Programmable personal fabricators are not just a prediction, they're a reality. The world of tomorrow can be glimpsed in tools available today. *Fab* tells the stories of these remarkable tools and their equally remarkable users around the world. It explains what can be made, and why, and how.

I first encountered the possibility of personal fabrication through the unexpectedly enthusiastic student response to a class that I teach at MIT, modestly titled "How To Make (almost) Anything." At MIT I direct the Center for Bits and Atoms. CBA comprises fifteen or so faculty from across campus: physicists, chemists, biologists, mathematicians, and mechanical and electrical engineers. They all, like me, never fit into the artificial separation of computer science from physical science.

The universe is literally as well as metaphorically a computer. Atoms, molecules, bacteria, and billiard balls can all store and transform information. Using the discrete language of computation rather than the continuous equations of calculus to describe the behavior of physical systems is not only leading to the practical development of new and more powerful kinds of information technologies, such as quantum computers, it's also leading to new kinds of insights into the nature of the universe itself, such as the long-term behavior of black holes. If the world is a computer, then the science of computing is really the science of science.

At the intersection of physical science and computer science, programs can process atoms as well as bits, digitizing fabrication in the same way that communications and computation were earlier digitized. Ultimately, this means that a programmable personal fabricator will be able to make anything, including itself, by assembling atoms. It will be a self-reproducing machine. That idea has been a long-standing science fiction staple for better or, sometimes, much worse.

In *Star Trek: The Next Generation*, the replicator is an essential plot element that is capable of making whatever is needed for each episode. It looks like an overgrown drinks dispenser, but it has the useful feature of being able to dispense anything. In theory, it does this by following stored instructions to put together subatomic particles to make atoms, atoms to make molecules, and molecules to make whatever you want. For Captain Picard, that was frequently a steaming mug of his preferred tea, obtained from the replicator with the command "Tea, Earl Grey, hot."

The less fortunate Arthur Dent in the *Hitchhiker's Guide to the Galaxy* had to contend with the infamous Nutri-Matic machine to obtain his cup of tea. Rather than storing the molecular specification in advance, the Nutri-Matic attempted to personalize Arthur's beverage by performing a spectroscopic analysis of his metabolism, and then probing the taste centers in his brain. As with Captain Picard's tea, Arthur's drink is synthesized by assembling its molecular constituents. However, in the Nutri-Matic's case the inevitable result was a plastic cup filled with a liquid that was almost, but not quite, entirely unlike tea.

None of this violates any physical laws, and in fact such atomic-scale programmable assembly is already possible in the lab today (as long as your tastes don't run to anything much larger than a few atoms).

To develop real working personal fabricators that can operate on a larger scale, my colleagues at MIT and I assembled an array of machines to make the machines that make machines. These tools used supersonic jets of water, or powerful lasers, or microscopic beams of atoms to make—well, almost anything. The problem we quickly ran into was that it would take a lifetime of classes for students to master all of the tools, and even then the students would get little practical experience in combining these tools to create complete working systems. So, we thought, why not offer a single-semester course that would provide a hands-on introduction to all the machines?

In 1998 we tried teaching "How To Make (almost) Anything" for the first time. The course was aimed at the small group of advanced students who would be using these tools in their research. Imagine our surprise, then, when a hundred or so students showed up for a class that could hold only ten. They weren't the ones we expected, either; there were as many artists and architects as engineers. And student after student said something along the lines of "All my life I've been waiting to take a class like this," or "I'll do anything to get into this class." Then they'd quietly ask, "This seems to be too useful for a place like MIT—are you really allowed to teach it here?"

Students don't usually behave that way. Something had to be wrong with this class, or with all the other classes I taught. I began to suspect the latter.

The overwhelming interest from students with relatively little technical experience (for MIT) was only the first surprise. The next was the reason why they wanted to take the class. Virtually no one was doing this for research. Instead, they were motivated by the desire to make things they'd always wanted, but that didn't exist. These ranged from practical (an alarm clock that needs to be wrestled into turning off), to fanciful (a Web browser for parrots), to profoundly quirky (a portable personal space for screaming). Their inspiration wasn't professional; it was personal. The goal was not to publish a paper, or file a patent, or market a product. Rather, their motivation was their own pleasure in making and using their inventions.

The third surprise was what these students managed to accomplish. Starting out with skills more suited to arts and crafts than advanced engineering, they routinely and single-handedly managed to design and build complete functioning systems. Doing this entailed creating both the physical form—mastering the use of computer-controlled tools that produce three-dimensional shapes by adding or removing material—and the logical function—designing and building circuits containing embedded computer chips interfaced with input and output devices. In an industrial setting these tasks are distributed over whole teams of people who conceive, design, and produce a product. No one member of such a team could do all of this, and even if they could, they wouldn't: personal screaming technology is unlikely to emerge as a product plan from a marketing meeting (even if the participants might secretly long for it).

The final surprise was how these students learned to do what they did: the class turned out to be something of an intellectual pyramid scheme. Just as a typical working engineer would not have the design and manufacturing skills to personally produce one of these projects, no single curriculum or teacher could cover the needs of such a heterogeneous group of people and machines. Instead, the learning process was driven by the demand for, rather than supply of, knowledge. Once students mastered a new capability, such as waterjet cutting or microcontroller programming, they had a near-evangelical interest in showing others how to use it. As students needed new skills for their projects they would learn them from their peers and then in turn pass them on. Along the way, they would leave behind extensive tutorial material that they assembled as they worked. This phase might last a month or so, after which they were so busy using the tools that they couldn't be bothered to document anything, but by then others had taken their place. This process can be thought of as a "just-in-time" educational model, teaching on demand, rather than the more traditional "just-in-case" model that covers a curriculum fixed in advance in the hopes that it will include something that will later be useful.

These surprises have recurred with such certainty year after year that I began to realize that these students were doing much more than taking a class; they were inventing a new physical notion of literacy. The common understanding of "literacy" has narrowed down to reading and writing, but when the term emerged in the Renaissance it had a much broader meaning as a mastery of the available means of expression. However, physical fabrication was thrown out as an "illiberal art," pursued for mere commercial gain. These students were correcting a historical error, using millions of dollars' worth of machinery for technological expression every bit as eloquent as a sonnet or a painting.

Today there aren't many places where these kinds of tools are available for play rather than work, but their capabilities will be integrated into accessible and affordable consumer versions. Such a future really represents a return to our industrial roots, before art was separated from artisans, when production was done for individuals rather than masses. Life without the infrastructure we take for granted today required invention as a matter of survival rather than as a specialized profession. The design, production, and use of engineered artifacts—agricultural implements, housewares, weapons, and armor—all took place locally. The purpose of bringing tool-making back into the home is not to recreate the hardships of frontier living, just as it's not to run personal-scream-container production lines out of the family room. Rather, it's to put control of the creation of technology back in the hands of its users.

The analogy between the personalization of fabrication and computation is close, and instructive. Remember that mainframes were behemoths; in 1949 *Popular Mechanics* famously forecast that "Computers in the future may weigh no more than 1.5 tons." The Digital Equipment Corporation (DEC) pioneered computers that were the size of desks rather than rooms. They called these "Programmed Data Processors" (PDPs) rather than computers because the market for computers was seen as being too small to be viable. But over time this class of devices came to be called *minicomputers*. A 1964 DEC ad proudly proclaimed, "Now you can own the PDP–5 computer for what a core memory alone used to cost: $27,000." That's a lot more than a PC costs today, but a lot less than the price of a mainframe. Minicomputers were thus accessible to small groups of users rather than just large corporations, and consequently their uses migrated from meeting the needs of corporations to satisfying individuals. PDPs did eventually shrink down so that they could fit on a desk, but they didn't end up there, because the engineers developing them didn't see why nonengineers would want them. DEC's president, Ken Olsen, said in 1977 that "there is no reason for any individual to have a computer in their home." PCs are now in the home; DEC is now defunct.

The adoption of PCs was driven by "killer apps," applications that were so compelling they motivated people to buy the systems to run them. The classic killer app was the original spreadsheet, VisiCalc, which in 1979 turned the Apple II from a hobbyist's toy to a serious business tool, and helped propel IBM into the PC business. VisiCalc's successor, Lotus 1–2–3, did the same in 1983 for IBM's PCs. The machine tools that the students taking "How To Make (almost) Anything" use today are much like mainframes, filling rooms and costing hundreds of thousands of dollars. But despite their size and expense, they've been adequate to show, year in and year out, that the killer app for personal fabrication is fulfilling individual desires rather than merely meeting mass-market needs. For one student this meant putting a parrot online; for another an alarm clock that got her up in the morning. None of the students needed to convince anyone else of the value of their ideas; they just created them themselves.

The invention that made PCs possible was integrated circuits, leading up to the development of microprocessors that put the heart of a computer onto a single silicon chip. The invention that promises to put the capabilities of a roomful of machine tools onto a desktop is the printing of functional materials. A printer's ink-jet cartridge contains reservoirs of cyan, magenta, yellow, and black inks; by precisely placing tiny drops it's possible to mix these to produce what appear to be perfect reproductions of any image. In the research lab today, there are similar inks that can be used to print insulators, conductors, and semiconductors to make circuits, as well as structural materials that can be deposited to make three-dimensional shapes. The integration of these functional materials into a printing cartridge will make possible faithful reproductions of arbitrary objects as well as images. At MIT we have a joke that we now take seriously: a student working on this project can graduate when their thesis can walk out of the printer. In other words, along with printing the text of the document, the printer must produce the means for the thesis to move.

Ultimately, rather than relying on a printer to place droplets of material, the logic for assembling an object will be built into the materials themselves. This is exactly how our bodies are made; a molecular machine called the ribosome translates instructions from genes into the series of steps required to assemble all of the proteins in our bodies out of the twenty amino acids. The discovery of building with logic is actually a few billion years old; it's fundamental to the emergence of life. Current research is now seeking to do the same with functional materials, creating a fundamentally digital fabrication process based on programming the assembly of microscopic building blocks. This mechanism will be embodied in personal fabricators fed by such structured materials. Much as a machine today might need supplies of air, water, and electricity, a digital personal fabricator will use as raw feedstocks streams of conductors, semiconductors, and insulators.

Unlike machines of today, though, but just like a child's building blocks, personal fabricators will also be able to disassemble something and sort its constituents, because the assembled objects are constructed from a fixed set of parts. The inverse of digital fabrication is digital recycling. An object built with digital materials can contain enough information to describe its construction, and hence its deconstruction, so that an assembler can run in reverse to take it apart and reuse its raw materials.

We're now on the threshold of a digital revolution in fabrication. The earlier revolutions in digitizing communications and computation allowed equipment made from unreliable components to reliably send messages and perform computations; the digitization of fabrication will allow perfect macroscopic objects to be made out of imperfect microscopic components, by correcting errors in the assembly of their constituents.

Return now to the mainframe analogy. The essential step between mainframes and PCs was minicomputers, and a similar sequence is happening along the way to personal fabrication. It's possible to approximate the end point of that evolution today with a few thousand dollars of equipment on a desktop, because engineering in space and time has become cheap.

First the space part. An inexpensive CD player places its read head with a resolution of a millionth of a meter, a micron. Combining this metrology with a desktop computer-controlled milling machine enables it to move a cutting tool in three dimensions with that same resolution, patterning shapes by removing material with tolerances approaching the limits of human perception. For example, it can cut out circuit boards with features as fine as the tiniest components.

Then the time part. A one-dollar embedded computer chip can operate faster than a millionth of a second, a microsecond. This is fast enough to use software to perform functions that traditionally required custom hardware, such as generating communications signals and controlling displays. It is possible to program one such chip to take on the functions of many different kinds of circuits.

The increasing accessibility of space and time means that a relatively modest facility (on the scale of the MIT class) can be used to create physical forms as fine as microns and program logical functions as fast as microseconds. Such a lab needs more complex consumables than the ink required by a printer, including copper-clad boards to make circuits and computer chips to embed into projects. But, as the students found at MIT, these capabilities can be combined to create complete functioning systems. The end result is much the same as what an integrated PF will be able to do, allowing technological invention by end users.

Minicomputers showed how PCs would ultimately be used, from word processing to e-mail to the Internet, long before technological development caught up to make them cheap enough and simple enough for wider adoption. Struck by this analogy, I wondered if it would be possible to deploy proto–personal fabricators in order to learn now about how they'll be used instead of waiting for all of the research to be completed.

This thought led to the launch of a project to create field "fab labs" for exploring the implications and applications of personal fabrication in those parts of the planet that don't get to go to MIT. As you wish, "fab lab" can mean a lab for fabrication, or simply a fabulous laboratory. Just as a minicomputer combined components—the processor, the tape drive, the keypunch, and so forth—that were originally housed in separate cabinets, a fab lab is a collection of commercially available machines and parts linked by software and processes we developed for making things. The first fab labs have a laser cutter to cut out two-dimensional shapes that can be assembled into three-dimensional structures, a sign cutter that uses a computer-controlled knife to plot flexible electrical connections and antennas, a milling machine that moves a rotating cutting tool in three dimensions to make circuit boards and precision parts, and the tools for programming tiny high-speed microcontrollers to embed logic. A bit like the original PDP minicomputers, all of this could be called a Programmed Materials Processor. This is not a static configuration; the intention over time is to replace parts of the fab lab with parts made in the fab lab, until eventually the labs themselves are self-reproducing.

The National Science Foundation (NSF) provided the seed funding for fab labs through its support of the Center for Bits and Atoms (CBA). NSF expects research activities that it funds on the scale of CBA to have an educational outreach component, which all too often is limited to teaching some classes at a local school, or creating a Web site describing the research. Instead, my CBA colleagues and our NSF counterparts agreed to try equipping ordinary people to actually *do* what we're studying at MIT instead of just talking about it. It's possible in the fab labs to do work that not too long ago required the resources of a place like MIT (an observation that's not lost on me on days when the administrative overhead of working at a place like MIT is particularly onerous).

Starting in 2002, the first fab labs went to rural India, Costa Rica, northern Norway, inner-city Boston, and Ghana. The equipment and supplies for each site initially cost about twenty thousand dollars. Knowing that that cost will come down as the technology progresses, the first fab labs weren't meant to be economically self-sustaining. One of the first surprises from the field was the demand for duplicating the labs even at that cost.

In keeping with the fab lab project's goal of discovering which tools and processes would be most useful in the field, we started setting up these labs long before we knew how best to do it. The response in the field was as immediate as it had been at MIT. We ended up working in so many far-flung locations because we found a demand for these capabilities around the world that was every bit as strong as that around campus. In the village of Pabal in western India, there was interest in using the lab to develop measurement devices for applications ranging from milk safety to agricultural engine efficiency. In Bithoor, on the bank of the Ganges, local women wanted to do three-dimensional scanning and printing of the carved wooden blocks used for chikan, a local kind of embroidery. Sami herders in the Lyngen Alps of northern Norway wanted wireless networks and animal tags so that their data could be as nomadic as their animals. People in Ghana wanted to create machines directly powered from their abundant sunlight instead of scarce electricity. Children in inner-city Boston used their fab lab to turn scrap material into sellable jewelry.

For all the attention to the "digital divide" in access to computers between developed and developing countries, these recurring examples suggest that there is an even more significant divide in access to tools for fabrication and instrumentation. Desktop computers are of little use in places that don't have desks; all too often they sit idle in isolated rooms built by aid agencies to house them. Appropriate computing requires the means to make, measure, and modify the physical world of atoms as well as the virtual world of bits. And instead of bringing information technology (IT) to the masses, fab labs show that it's possible to bring the tools for IT development, in order to develop and produce local technological solutions to local problems.

There's been a long-standing bias that technology's role in global development mirrors its own history, progressing from low- to high-tech. The fab lab experience suggests instead that some of the least developed parts of the world need some of the most advanced technologies. That observation has led me to spend some head-spinning days in Washington, DC, going from the World Bank to the National Academy of Sciences to Capitol Hill to the Pentagon, having essentially the same meeting at each place. Fab labs challenge assumptions that are fundamental to each of these institutions. Instead of spending vast sums to send computers around the world, it's possible to send the means to make them. Instead of trying to interest kids in science as received knowledge, it's possible to equip them to *do* science, giving them both the knowledge and the tools to discover it. Instead of building better bombs, emerging technology can help build better communities.

The problem I found on these trips was that none of these institutions knew how to pay for this kind of work; there isn't a Pentagon Office of Advanced Technologies for Avoiding Wars. Financing personal fabrication in underserved communities is too directed a goal for traditional basic research funding, and too speculative for conventional aid organizations or donors. The closest precedent is microcredit lending, which provides small loans to help support financial cooperatives, typically run by women, in developing countries. The loans are used to acquire an asset such as a cell phone that can then be used to generate income. But that model doesn't help when the financing is for an invention. What's needed are the skills of a good venture capitalist rather than a banker. That's not an oxymoron; the best venture capitalists add value to their investments by helping shepherd, prune, and protect ideas, build operational teams, and develop business models.

The historical parallel between personal computation and personal fabrication provides a guide to what those business models might look like. Commercial software was first written by and for big companies, because only they could afford the mainframe computers needed to run it. When PCs came along anyone could become a software developer, but a big company was still required to develop and distribute big programs, notably the operating systems used to run other programs.

Finally, the technical engineering of computer networks combined with the social engineering of human networks allowed distributed teams of individual developers to collaborate on the creation of the most complex software.

Programmers write source code that people can understand, which gets turned into executable code that computers can understand. Commercial companies have protected the former and distributed the latter to their customers. But individuals who share the source code that they write can collaborate in ad hoc groups, which might never meet physically, on the creation of programs that are larger than any one of the members could write alone. The Linux operating system is built out of such "open source" software. Much like the way science progresses by researchers building on one another's publications, a programmer can make available a piece of code that might then get taken up and improved by someone on the opposite end of the earth.

In a world of open-source software, ownership of neither computers nor code alone provides the basis for a proprietary business model; what's left is the value added to them by creating content and delivering services. Profitable old and new computer companies are making money from freely available software in just this way: by charging for their role in solving problems.

Similarly, possession of the means for industrial production has long been the dividing line between workers and owners. But if those means are easily acquired, and designs freely shared, then hardware is likely to follow the evolution of software. Like its software counterpart, open-source hardware is starting with simple fabrication functions, while nipping at the heels of complacent companies that don't believe that personal fabrication "toys" can do the work of their "real" machines. That boundary will recede until today's marketplace evolves into a continuum from creators to consumers, servicing markets ranging from one to one billion.

That transition has already happened in the case of two-dimensional printers, the immediate predecessor to personal fabricators. High-quality printing, once solely a commercial service, came into the home via laser printers. The most important attribute of an industrial printing press is its throughput, which is the number of pages it can produce per minute. Laser printers started going down this technological scaling curve with the development of ever-faster printers needed for serious on-demand commercial printing. Within Hewlett-Packard, a competing group of engineers had the idea that squirting individual drops of ink could make beautiful images more cheaply than transferring toner onto sheets. Ink-jet printing would be slower than laser printing, but they reasoned that for a printer in the home, quality mattered much more than speed. This was such a heretical idea that this group decamped from HP's headquarters in Palo Alto and set up shop out of sight in Corvallis, Oregon. The rest is business history; the relative cost of producing and selling ink-jet cartridges has been about the closest a company has come to legally printing money. Ink-jet printing hasn't replaced commercial printing; it's created an entirely new—and enormous—market driven by quality and access rather than speed.

Similarly, the emerging personal fabrication tools I've been describing are intended for personal rather than mass production. Their development was originally driven in industry by the need to quickly create prototypes of products to catch errors before they became much more expensive to correct in production. Machine-tool shows relegate such rapid-prototyping machines to a sleepy corner away from the giant cutting, stamping, and molding tools that are at the top of the machine food chain. But if the market is just one person, then the prototype *is* the product. The big machines will continue to mass-produce things used in large quantities; nuts and bolts are valuable because they're identical rather than unique. But little machines will custom-make the products that depend on differences, the kinds of things being made in the fab class and fab labs.

The biggest impediment to personal fabrication is not technical; it's already possible to effectively do it. And it's not training; the just-in-time peer-to-peer project-based model works as well in the field as at

MIT. Rather, the biggest limitation is simply the lack of knowledge that this is even possible. Hence this book.

Fab tells the stories of pioneering personal fabrication users, and the tools they're using. Because both are so striking, I've interwoven their stories in pairs of chapters that explore emerging applications and the processes that make them possible. The not-so-hidden agenda is to describe not only who is doing what but also how, providing introductions to the tools that are similar to the orientations we give at MIT and in the field. These stop just short of hands-on training; a final section gives enough detail on the products, programs, and processes used to duplicate what's shown in the book.

Throughout *Fab* I use what are known as "hello world" examples. In 1978, the instruction manual for the then new C programming language written at Bell Labs used as an example a simple program that printed out the words "hello world." This is more exciting than it sounds, because it requires an understanding of how to write a rudimentary program, compile it into computer code, and cause the program to print the text. "Hello world" programs have since become staples for introducing new computer languages; the Association for Computing Machinery currently lists 204 examples for languages from A+ to zsh.

The difference between those "hello world" programs and the examples I give in this book is that mine arrange atoms as well as bits, moving material as well as instructions. But the principle is the same: show the minimum specification needed to get each of the tools to demonstrate its successful operation. Taken together, these examples provide a fairly complete picture of the means for almost anyone to make almost anything.

My hope is that *Fab* will inspire more people to start creating their own technological futures. We've had a digital revolution, but we don't need to keep having it. Personal fabrication will bring the programmability of the digital worlds we've invented to the physical world we inhabit. While armies of entrepreneurs, engineers, and pundits search for the next killer computer application, the biggest thing of all coming in computing lies quite literally out of the box, in making the box.

. . . Almost Anything

Kelly

The first year that we taught "How To Make (almost) Anything" at MIT, Kelly Dobson was the star student. She was an artist, with lots of ideas for projects but little electronics background. Like the other students in the class, she applied the skills she had learned to her semester project. Unlike the others, though, she chose to focus on the problem of needing to scream at an inconvenient time. The result was Kelly's ScreamBody, which looks like a fuzzy backpack worn in front rather than in back.

As Kelly describes it during a demonstration: "Have you ever really wanted to scream? But you can't, because you're at work, or in any number of situations where it's just not appropriate? Well, ScreamBody is a portable personal space for screaming. As a person screams into ScreamBody their scream is silenced, but it is also recorded for later release where, when, and how they choose."

There's a familiar sequence of reactions when people hear this. First, shock: Is she really making a machine to save screams? Then amusement, when they see that it actually works. And then more often than

ScreamBody

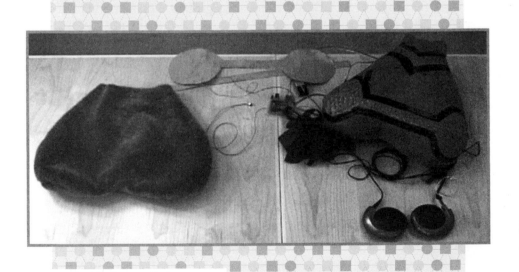

Inside ScreamBody

not they express a desire to get one of their own. Finally comes the appreciation of what Kelly has in fact accomplished in making ScreamBody work.

The first thing she did was design a circuit to save screams, then she built that into a circuit board. This in turn required programming an embedded computer to control the sound recording and playback. Next she developed sensors for the screamer to interact with the ScreamBody, and finally assembled the apparatus into a comfortable, wearable package that could silence and later emit the scream.

In a conventional corporation, a strategy group would come up with the ScreamBody's specifications, electrical engineers would design the circuits, computer scientists would program the chips, mechanical engineers would make the structures, industrial designers would assemble everything, industrial engineers would plan the production, and then marketers would try to find someone to buy it. The only thing that Kelly finds remarkable about doing all of this herself is that anyone finds it remarkable. For Kelly, literacy means knowing how to communicate by using all the representations that information can take. She sees the design of circuits and packaging as aspects of personal expression, not product development. She didn't design ScreamBody to fill a market need; she did it because she wanted one.

Unlike personal computers, or personal digital assistants, which are designed by teams of engineers to meet the common needs of the largest possible market, Kelly's personal screaming assistant is targeted at a market base of one user: Kelly. What makes her exceptional is that, like most everyone else, she's not average. For her, appropriate technology helps her scream; for others it might help them to grow food or play games or customize jewelry. Individual needs are unlikely to be met by products aimed at mass markets. A truly personal computing device is by definition not mass-produced, or even mass-customized; it is personally designed.

Meejin

In the years since that first class I've been delighted to continue to find that students—ranging from undergrads to faculty members—have wildly different and equally quirky visions of personally appropriate technologies. Kelly was followed by Meejin Yoon, a new professor in the architecture department who like Kelly had no technical fabrication experience. Meejin took the class because she wanted to be able to make her designs herself.

Meejin was struck by the ways that technology is encroaching on our personal space, and wondered if instead it could help to define and protect it. She expressed that thought in the Defensible Dress, which was inspired by the approach of the porcupine and blowfish to protecting their personal space. Her device looks like a fringed flapper dress, but the fringes are stiff piano wire connected to the dress by motors controlled by proximity sensors. When someone comes within a distance determined by the wearer, the wires spring out to announce and defend the space around it. Countless commercial wearable computers are aimed at maximizing communication; this is one that aims to minimize it.

Defensive dressing

Shelly

After Meejin came Shelly Levy-Tzedek, a budding biologist who was unfamiliar with how to build with things bigger than molecules. She chose to focus on one of the most pressing practical concerns for an MIT student: waking up. Alarm clocks now on the market are too easily defeated by sleep-deprived scientists.

Shelly decided to make the match more even by raising the IQ of the alarm clock, so that it could present a challenge to the would-be sleeper. She came up with an alarming-looking clock; instead of simply turning off a button or activating a snooze mode, Shelly's clock demands that the user grab glowing protuberances in the order in which they randomly flash. That's a challenging enough game when

An alarming clock

I have re-designed my board to include two PIC chips - one "main" pic, controlling the time and alarm setting and comparing the two to decide whether to go off, and one "game" pic, controlling the LEDs and capacitive sensing (touch pads). next I'll cut it out on the HAAS machine.

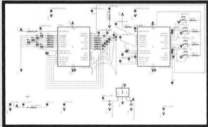 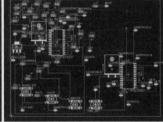

in the very last minute I decided to add extra "decoy" rods that are left with the lights turned on at all times.

Fab log

you're awake, but it becomes fiendishly difficult when you're fumbling about half-awake in the dark. When Shelly displayed the clock at a reception for fab class projects, the universal reaction was "Where can I get one?"

The path that led Shelly to the clock was as interesting as the clock itself. As she worked she kept a technical log on a Web page that recorded her early ideas, emerging engineering files, and observations about what did and didn't work. The students in the class used these kinds of pages to exchange designs, learn about tricks and traps, and share their thoughts about problem solving (as well as one another). These evolving pages grew into the instructional material for the students, as well as a lasting record for their successors.

Dalia

Some of those successors were far from MIT, institutionally as well as physically. An inner-city Boston community center, the South End Technology Center (SETC), served as a fab lab laboratory to try out small versions of the big machines used in the fab class at MIT: a laser cutter for making two-dimensional shapes and parts for three-dimensional structures, a sign cutter for plotting flexible circuits as well as graphics, a milling machine for making precision structures and circuit boards, and tools for assembling and programming circuits.

An early interest among kids at SETC was using these tools to create custom controllers for computer games, using the same kind of proximity sensors that Shelly had used in her clock. To make these circuits we used "surface-mount" components, placed on top of rather than

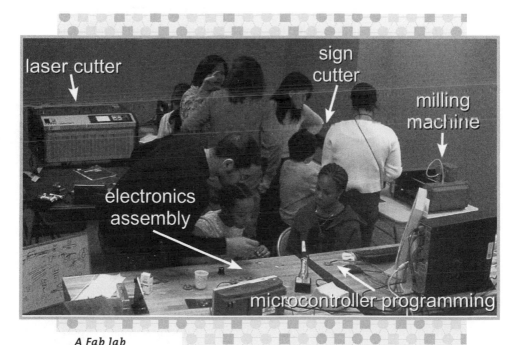

A Fab lab

going through a circuit board. This makes the circuit board itself smaller and easier to fabricate, but it requires more skill to assemble the parts. As we developed the fab lab, the plan was to first teach a few interested MIT students how to create the circuit boards, and then have them show the kids at the community center. But when I prepared to show a grad student, Amon Millner, how to do that, 11-year-old Dalia Williams showed up. Dalia thought the idea sounded pretty cool, so she shoved Amon aside and announced that she was going to make the board instead of him.

Dalia appeared to be entirely out of control; parts were flying everywhere. And she wasn't entirely clear on how it all worked; she and a friend thought the solder they were using was called "salsa." But Dalia stuck with the project, practicing and persisting until she had all of the parts on the board. I was perhaps the only person there who was surprised that it worked the first time we powered it up.

As with Kelly at MIT, other girls followed Dalia's footsteps. First the novelty of the tools attracts them, then the joy of using them for personal expression hooks them. For these girls, the tangibility of what they make is far more compelling than just clicking at programs on a computer screen.

The girls at SETC are also finding that the things they make can be commercially as well as personally rewarding. A group of them set up the fab lab tools on a nearby street corner and held a high-tech craft sale of on-demand creations, making about a hundred dollars in an afternoon. This was a life-transforming event for kids growing up in inner-city Boston. Beyond their individual projects, they're discovering that they can create one of the most valuable things of all: a job.

The best description of the accomplishments of Dalia and her peers came accidentally from an errant newscaster. He was reporting on a workshop run by my colleague Mitch Resnick, aimed at engaging girls from underserved communities in scientific thinking through experimenting with applications of feedback control systems.

A diary security system

The girls used computing "bricks" that contained circuits similar to the one Dalia built. One of the girls chose to make a diary security system that would take a picture of anyone (like her brother) who came near her diary. To do this she connected a sensor and a camera to the embedded computer and wrote a program to control the camera with the sensor. The reporter generously observed that "contrary to studies, girls can do anything, even science!"

The comment is of course egregious, but the conclusion is not. Girls (and boys) increasingly *can* do and make anything. Their futures are literally in their own hands. Their needs, and thus their projects, aren't the same as those of ordinary engineers, but the tools and skills they're using are much the same. In mastering these emerging means for personal fabrication, they're helping bring individual expression back into mass manufacturing, the subject of the next section.

The Past

Once upon a time, education, industry, and art were integrated in the work of a village artisan. By the time that I went to school, college-bound kids like me had to sit in rather sterile classrooms, while the kids taking up trades got to go to a vocational school that had all the cool stuff—machine tools, welders, electronic test equipment, and the like. At the time, this split seemed vaguely punitive for someone like me. I couldn't understand why an interest in making things was taken as a sign of lesser intelligence. It wasn't until I became an MIT professor and had an official excuse to buy those kinds of tools that I realized the problem predates my high school by quite a bit. The Kellys and Meejins and Shellys and Dalias of the world are correcting a historical error that dates back to the Renaissance.

In the fifteenth and sixteenth centuries, stiff paintings of religious subjects gave way to much more diverse images of breathtaking realism and luminosity. The advances that made this possible were first of all technical, including the introduction of (originally Flemish) oil paints and the development of mathematical techniques for establishing perspective. Oil-based paints dried more slowly, allowing more complex brushstrokes; the intensity of the colors led to thinner layers that better reflected light; and the viscosity of the paints improved their coverage on wood and canvas. At the same time, vision studies led to geometrical solutions for the problem of how to best project a three-dimensional scene onto a two-dimensional surface. These techniques were created by and for artists; Leonardo da Vinci, for example, was continually experimenting with new paints.

Improvements in paints and perspective would not have had the impact they did if not for the simultaneous development of artists to use them. Your average Middle Ages painter worked as an artisan in a guild, with the same (lack of) creative control as a carpenter. An aspiring painter would progress from apprentice to journeyman to master, finally gaining full admission to the guild by producing a "masterpiece." The real mastery of the guilds, however, was of the marketplace: they were very effective monopolists, controlling both the supply of and standards for skilled laborers. The work itself was done to detailed specifications drawn up by, say, a church that wanted an altarpiece illustrated with a particular scene.

The guild system began to break down under both the fragmentation of the crafts into increasingly narrow specialties and their aggregation into workshops that could produce increasingly complex, complete artifacts (and which formed the seeds for the later development of factories). But creative individuals were able to escape from the guild system because of another invention: customers. Artisans became artists when a population emerged that had both the discretionary income and the intellectual interest to acquire art.

Led by merchant families, most notably the Medicis in Florence, and the occasional municipality or pre-Enlightened monarch, a community of patrons began to emerge that bought art from and for individuals rather than (literally) dogmatic institutions. Michelangelo and Leonardo da Vinci started their careers as apprentices but ended up valued for their personal expression rather than their occupational productivity. Leonardo da Vinci ultimately represented just himself. He was not the CEO of a nascent da Vinci Industries, with a fiduciary responsibility to its shareholders to maximize the brand's value (which is a good thing, otherwise his legacy might instead have been a line of Mona Lisa and Luigi dolls).

As remarkable as these new artists and their new materials and methods for painting were, their subject material was more significant still. Paintings began to represent the world of people rather than that of myths or gods. This human focus both reflected and shaped the defining intellectual movement of the Renaissance: humanism. What liberal arts students now study as the humanities emerged in the Renaissance as an expression of a growing sense of mastery by (selected) people over their world.

In Italy, humanism grew in part out of a fifteenth-century attempt to understand Roman and Greek ruins, both architectural and intellectual, an effort that today would be called reverse engineering. While much ancient knowledge and experience had been lost in the intervening centuries, the existing record of ruins and ancient documents provided a handy template for reconstructing a society that had worked pretty well for the Romans (other than the decline-and-fall issue). This enterprise benefited considerably when the Ottomans conquered Constantinople in 1453, freeing up a ready supply of fleeing Greek scholars. When they sought refuge in Italy they brought with them writings and knowledge that had long been lost to the West.

A second ingredient in the emergence of humanism arose as a reaction against the strictures and scriptures of the prevalent monastic, ecclesiastical seat of advanced education. While nominally still religiously observant, the growing urban mercantile economy and rule by civic authority brought a need for training future secular leaders with relevant skills. The human part of *humanism* comes from *studia humanitatis* ("studies of humanity"), referring to this shift in focus from immortal to mortal subjects, and was associated with a growing interest in how people were reflected in everything from portraiture to pedagogy.

These threads of humanism came together with the dusting off (sometimes quite literally) of a curriculum based around the four-part quadrivium (geometry, arithmetic, astronomy, music) and the three-part trivium (grammar, logic, rhetoric). These Latin terms refer back to four- and three-way road intersections, the latter notable as a place where people would congregate and share knowledge that naturally came to be known as trivial, or trivia. The trivium and the quadrivium together make up the seven "liberal arts." Both of these words warrant comment. "Liberal" in this sense is not the opposite of "conservative"; it referred to the liberation that the study of these subjects was thought to bring. And "art" did not mean just creative expression; it meant much more broadly the mastery that was developed over each of these domains. Liberal arts originally has this rather rousing meaning as a mastery over the means for personal liberation. They're now associated with academic study that is remote from applications, but they emerged in the Renaissance as a humanist pathway to power.

In 1513 Niccolò Machiavelli wrote *The Prince*, the classic (and infamous) guide to governance, on how to use rhetoric to win friends and manipulate people. Renaissance social engineering also gave birth to the concept of utopia, if not the reality. The term first appeared in a book of that name, written by Sir Thomas More in 1516; his utopian vision was very much a humanist paradise, governed by reason and based on a belief in the power of ideas. It was against this backdrop of the growing recognition of the practical importance of language and reasoning that familiarity with the liberal arts emerged as a modern notion of literacy. These skills became an expectation of any active participant in civil society.

Unfortunately, the ability to make things as well as ideas didn't make the cut; that was relegated to the *artes illiberales*, the "illiberal arts," that one pursued for mere economic gain. With art separated from artisans, the remaining fabrication skills were considered just mechanical production. This artificial division led to the invention of unskilled labor in the Industrial Revolution.

As with the revolution in painting in the Renaissance, this transition in industry was triggered in part by advances in materials, in this case the use of iron and steel, which in turn both led to and benefited from the development of steam power. These developments in materials and power made possible modern machinery, most notably mechanized looms. These could produce much more cloth than traditional artisans could (from 50,000 pieces in 1770 England to 400,000 pieces in 1800), and thus could clothe many more people (from 8.3 million people in 1770 England to 14.2 million in 1821). Newly unemployed craft workers crowded into growing cities to seek employment operating the machines that would replace not only the jobs but also the skills of still more workers. Unintended consequences of this shift included a layer of black smoke covering earth and sky, generated from burning coal in the factories, and the epidemics of cholera, smallpox, typhoid, typhus, and tuberculosis that followed from packing people around the factories.

This new division of labor between people and machines became explicit with Joseph-Marie Jacquard's invention of the programmable loom, first demonstrated in Paris in 1801. He introduced an attachment that could read instructions on punched cards (more reliably than Florida's voters) to control the selection of shuttles containing colored threads and thereby program patterns into fabric.

Because the looms could now follow instructions, their operators no longer needed to. The job of the weaver was reduced to making sure that the loom was supplied with thread and cards. Lyon's silk weavers, threatened by this challenge to their livelihood, rather reasonably destroyed Jacquard's looms. But the looms won; commercial weaving turned from a skilled craft into menial labor.

The invention of industrial automation meant that a single machine could now make many things, but it also meant that a single worker who used to do many things now did only one. Thinking about how to make things had became the business of specialized engineers; the Ecole Polytechnique was set up in France in 1794 to train them, and in Britain there was an unsuccessful attempt to forbid the export of both its engineers and the machines they developed because of the perceived strategic importance of both.

Tellingly, in Britain, where the separation between art and artisans was furthest along, scientific progress suffered. The great acoustics discoveries of the nineteenth century occurred in France and Germany, where there was a lively exchange in workshops that made both musical and scientific instruments, rather than in England, where *handwerk* had become a pejorative term.

From there, the relative meaning of literacy diverged for machines, their designers, and their users. First, the machines. Around 1812, the mathematician Charles Babbage conceived that it would be possible to construct a machine to do the tedious job of calculating mathematical tables, and in 1823 he received government support to build his "Difference Engine." He failed to finish it (Babbage was also a pioneer in bad management), but the Difference Engine did produce one very useful output: the inspiration for the Analytical Engine. Babbage realized that an improved steam-powered engine could follow instructions on Jacquard's punched cards to perform arbitrary mathematical operations, and could change the operation by changing the cards rather than the machine. By the mid-1830s Babbage had failed to complete this new machine as well, limited by the timeless problems of underfunding and mismanagement, and by the available manufacturing technology that could not make parts with the complexity and tolerances that he needed.

Jacquard's punched cards reappeared in 1882 when Herman Hollerith, a lecturer at MIT who had worked for the U.S. Census Bureau as a statistician, sought a way to speed up the hand-tallying of the census. He realized that the holes in the cards could represent abstract information that could be recorded electrically. The result was the Hollerith Electric Tabulating System, which counted the 1890 census in a few months rather than the years that a hand tally would have required. The greater consequence of this work was the launch in 1896 of his Tabulating Machine Company, which in 1924 became IBM, the International Business Machines Corporation.

Information-bearing punched cards made machines more flexible in what they could do, but that didn't change anyone's notion of the nature of people versus machines. That challenge surfaced in an initially obscure paper published by the twenty-four-year-old Alan Turing in Cambridge in 1936. In "On Computable Numbers, with an Application to the Entscheidungsproblem," he tackled one of the greatest outstanding mathematical questions of his day, the *Entscheidungsproblem* ("decision problem") posed by the great mathematician David Hilbert in 1928: can there exist, at least in principle, a definite procedure to decide whether a given mathematical assertion is provable? This is the sort of thing that someone like Turing, employed in the business of proving things, might hope to be possible. His rather shocking answer was that it wasn't. Alonzo Church, who would become Turing's thesis adviser at Princeton, independently published the same conclusion in 1936, but Turing's approach was later considered by everyone (including Church) to be much more clever.

To make the notion of "procedure" explicit, Turing invented an abstract mechanism that he called an LCM, a *logical computing machine* (everyone else just called it a Turing machine). This device had a paper tape that could contain instructions and data, and a head that could move along the tape reading those instructions and interpreting them according to a fixed set of rules, and could then make new entries onto the tape. This sort of machine could follow a procedure to test the truth of a statement, but Turing was able to show that simple questions about the working of the machine itself, such as whether or not it eventually halts when given a particular set of instructions, cannot be answered short of just watching the machine run. This means that it might be possible to automate the solution of a particular problem, but that there cannot be an automated procedure to test when such an approach will succeed or fail.

This dramatic conclusion set a profound limit on what is knowable. The seemingly steady advance of the frontiers of knowledge had halted at a fundamentally unanswerable question: the undecidability of testing a mathematical statement. But Turing's solution contained an even greater consequence: He showed that the particular details of the design of the Turing machine didn't matter, because any one of them can emulate any other one by putting at the head of its tape a set of instructions describing how the other one works. For instance, a Mac Turing machine can use a PC Turing machine tape by starting it off with a PC specification written in Mac language. This insight, now called the Church-Turing thesis, is the key to machine literacy. Any machine that can emulate a Turing machine can solve the same problems as any other, because it is general enough to follow machine translation instructions. This property has since been shown to be shared by systems ranging from DNA molecules to bouncing billiard balls.

Turing's thoughts naturally turned to building such a universal computing device; in 1946 he wrote a "Proposal for Development in the Mathematics Division of an Automatic Computing Engine (ACE)" for the UK's National Physical Laboratory (NPL). Like Babbage, Turing proved himself to be better at proposing machines than building them, but those machines of course did get built by his successors. The story picks up across the Atlantic, where, shades of Babbage's math tables, the U.S. Army funded the construction of an all-electronic machine to be built with vacuum tubes to calculate artillery range tables. The ENIAC (Electronic Numerical Integrator and Computer) was publicly dedicated at the University of Pennsylvania in 1946. Its first calculations weren't range tables, though. They were something much more secret and explosive: mathematical models for the nuclear bomb effort at Los Alamos. Those calculations arrived via John von Neumann; getting him interested in computers was perhaps ENIAC's most important consequence.

Von Neumann is on the short list of the smartest people of the past century; those who knew him might say that he *is* the short list. He was a math wizard at Princeton, where he overlapped with Turing, and he was an influential government adviser. When von Neumann heard about the ENIAC through a chance encounter he planted himself at the University of Pennsylvania, recognizing how much more the ENIAC could do than calculate range tables. It was, after all, the first general-purpose programmable digital electronic computer.

It was also a rather clumsy first general-purpose programmable digital electronic computer. It tipped the scales at a svelte thirty tons, and for maximum operational speed it was programmed by plugboards that took days to rewire. It's charitable to even call it programmable. But his experience with this computer did lead von Neumann to propose to the Army Ordnance Department in 1945 construction of the EDVAC (Electronic Discrete Variable Computer), and in 1946 he elaborated on this idea in a memo, "Preliminary Discussion of the Logical Design of an Electronic Computing Instrument." He made the leap to propose that programs as well as data could be stored electronically, so that the function of a computer could be changed as quickly as its data. He proved to be a better manager than Babbage or Turing; the EDVAC was finished in 1952 (although the first stored-programs computers became operational at the universities of Manchester and Cambridge in 1949).

Having invented the modern architecture of computers, von Neumann then considered what might happen if computers could manipulate the physical world outside of them with the same agility as the digital world inside of them. He conceived of a "universal constructor," with a movable head like a Turing machine, but able to go beyond making marks to actually move material. Because such a thing was beyond the capabilities of the barely functional computers of his day, he studied it in a model world of "cellular automata," which is something like an enormous checkerboard with local rules for how checkers can be added, moved, and removed. Von Neumann used this model to demonstrate that the combination of a universal constructor and a universal computer had a remarkable property: self-reproduction. The computer could direct the constructor to copy both of them, including the program for the copy to make yet another copy of itself. This sounds very much like the essence of life, which is in fact what von Neumann spent the rest of his life studying. I'll return to this idea in "The Future" to look at the profound implications for fabrication of digital self-reproduction.

While von Neumann was thinking about the consequences of connecting a universal computer to machinery to make things, the first general-purpose programmable fabricator was actually being built at MIT. The Whirlwind computer was developed there in the Servomechanism Laboratory, starting in 1945 and demonstrated in 1951. Intended for the operation of flight simulators, the Whirlwind needed to respond to real-time inputs instead of executing programs submitted as batch jobs. To provide instantaneous output from the computer, the Whirlwind introduced displays screens. But if the computer could control a screen, that meant that it could control other things in real time. At the request of the air force, in 1952 the Whirlwind was connected to an industrial milling machine. The mechanical components for increasingly sophisticated aircraft were becoming too difficult for even the most skilled machinists to make. By having the Whirlwind control a milling machine, shapes were limited only by the expressive power of programs rather than by the manual dexterity of people. The machines of Babbage's day weren't up to making computers, but finally computers were capable of making machines. This in turn raised a new question: How could designers tell computers how to make machines?

The answer was the development of a new kind of programming language for doing what became known as computer-aided manufacturing (CAM) with numerically controlled (NC) machines. The first of these, Automatically Programmed Tools (APT), ran on the Whirlwind in 1955 and became available on IBM's 704 computer in 1958. It was a bit like the theoretical programs for a Turing machine that would specify how to move the read/write head along the abstract tape, but now the head was real, it could move in three dimensions, and there was a rotating cutting tool attached to it. APT is still in use, and is in fact one of the oldest active computer languages.

APT was a machine-centric representation: it described steps for the milling machine to follow, not results that a designer wanted. Real computer aid on the design side came from the next major computers at MIT, the TX–0 and TX–2. These were testbeds for computing with transistors instead of vacuum tubes, and sported a "light pen" that allowed the operator to draw directly on the display screen. In 1960 Ivan Sutherland, a precocious student supervised by Claude Shannon (inventor of the theory of information that forms the foundation of digital communications), used the combination of the TX–2 and the light pen to create the seminal "Sketchpad" program. Sketchpad let a designer sketch shapes, which the computer would then turn into precise geometrical figures. It was the first computer-aided design (CAD) program, and remains one of the most expressive ones.

The TX–2 begat Digital Equipment Corporation's PDP (Programmed Data Processor) line of computers, aimed at work groups rather than entire organizations. These brought the cost of computing down from one million dollars to one hundred thousand and then ten thousand dollars, sowing the seeds for truly personal computing and the PCs to come. The consequences of that personalization of computation have been historic. And limited.

Personal computers embody centuries of invention. They now allow a consumer to use a Web browser to buy most any product, from most anywhere. But online shopping is possible only if someone somewhere chooses to make and sell what's wanted. The technology may be new, but the economic model of mass production for mass markets dates back to the origin of the Industrial Revolution.

Unseen behind e-commerce sites are the computers that run industrial processes. Connecting those computers with customers makes possible what Stan Davis calls "mass customization," allowing a production line to, for example, cut clothes to someone's exact measurements, or assemble a car with a particular set of features. But these are still just choices made from among predetermined options. The real expressive power of machines that make things has remained firmly on the manufacturing rather than the consumer side.

Literacy has, if anything, regressed over time to the most minimal meaning of reading and writing words, rather than grown to encompass the expressive tools that have come along since the Renaissance. We're still living with the historical division of the liberal from the illiberal arts, with the belief that the only reasons to study fabrication are for pure art or profane commerce, rather than as a fundamental aspect of personal liberation.

The past few centuries have given us the personalization of expression, consumption, and computation. Now consider what would happen if the physical world outside computers was as malleable as the digital world inside computers. If ordinary people could personalize not just the content of computation but also its physical form. If mass customization lost the "mass" piece and become personal customization, with technology better reflecting the needs and wishes of its users because it's been developed by and for its users. If globalization gets replaced by localization.

The result would be a revolution that contains, rather than replaces, all of the prior revolutions. Industrial production would merge with personal expression, which would merge with digital design, to bring common sense and sensibility to the creation and application of advanced technologies. Just as accumulated experience has found democracy to work better than monarchy, this would be a future based on widespread access to the means for invention rather than one based on technocracy.

That will happen. I can say this so firmly because it is my favorite kind of prediction, one about the present. All of the technologies to personalize fabrication are working in the laboratory, and they are already appearing in very unusual but very real communities of users outside the laboratory. The stories of these technologies, and people, are the subject of the rest of this book.

Hard Ware

If the conception of universal programmed assemblers is one of the peaks of human invention, then the reality of the engineering software used in manufacturing today is somewhere in the pits. This sorry state of affairs can appear to present an insurmountable obstacle to an unwary would-be personal fabricator; however, the same kind of rapid-prototyping approach to hardware can also be applied to create software aimed at the needs of mortals rather than machines.

The intuitive drawing metaphor of Ivan Sutherland's original Sketchpad program has been replaced in state-of-the-art engineering software with obscure "user interfaces." That those words are even used to describe how computers appear and respond to their users indicates the problem: people, like printers, are considered to be a kind of computer peripheral requiring a compatible communication interface. There's been no compelling reason to make engineering software easy to use; these programs have been written by engineers, for engineers, who make a career out of using one of them. Here's what the on-screen controls for one high-end computer-aided design (CAD) package look like:

One's immediate reaction on being greeted by this screen is "Huh, what does that do?" After extensive experience, one's reaction remains "Huh, what does that do?" It's just not intuitive. It is powerful: this program could be used to design a jumbo jet, keep track of the relationships among all of the parts, simulate how they fit together, model how they respond to loads, control the machines to produce them, and order the required materials. But "ease of use" and "design jumbo jet" have not been compatible concepts.

The generality of a high-end CAD package comes at a high price in complexity. In a big package each of its functions is typically accessed in a different operating mode, which entails running a different program. One mode is used to design individual parts, another to place them in assemblies, another to render how they look, and still another to specify how they're made. The need to explicitly switch among these modes is an obstacle to working in the way that an idea develops; it forces invention into a fixed sequence. If you're in assembly-editing mode and have an idea about improving one of the components, you have to hold that idea while taking all of the steps required to switch back to the mode to edit that part.

Then there's the economic cost. These programs historically were bought to run on million-dollar computers, and were used to control equally expensive machines. Just as the software for a thousand-dollar PC costs about a hundred dollars, the packages for million-dollar mainframes cost tens to hundreds of thousands of dollars. Since these programs are so expensive, it's not possible to do something so rash as merely run them. Instead, another computer configured as a license server must approve use of the software, or a dreaded hardware "dongle" installed on the computer must authorize use of the software. If the license is not configured exactly right, trying to run the package yields helpful error messages such as:

No, it's not OK.

Once you make it through this gauntlet and actually do design something, the real fun begins: transferring files between programs. The output of a CAD program goes to a CAM (computer-aided manufacturing) program, to translate the design into instructions that a particular manufacturing machine can understand. Here's what happened when I tried that for one of the "Hello World" examples we'll see later in the book:

Helplessly staring at garbled conversions is a depressingly familiar experience as soon as more than one engineering program is involved.

This "H-ll W il l" was saved in one of the most common engineering formats, as a Drawing Interchange File, which, as a clue to its problem, is abbreviated as DXF. (Where did the X come from?) Take the simplest task for an engineering file format, specifying the coordinates of a point, X and Y and Z. In the DXF documentation, one helpfully finds the right code to use to specify this in a DXF file:

```
|————|————————————————————————|
| 10       | Primary X coordinate (start point of a Line or    |
|          | Text entity, center of a Circle, etc.)            |
|————|————————————————————————|
```

Of course, if that doesn't suit, there are eight other kinds of X coordinates:

```
|————|————————————————————————|
| 11–18    | Other X coordinates                               |
|————|————————————————————————|
```

However, since the available evidence supports the belief that we live in a three-dimensional universe, just one kind of X coordinate should be adequate for anyone other than a theoretical particle physicist. That is, unless the X coordinate isn't just a regular X coordinate but rather is an X coordinate for an "extended entity," in which case a different code is used:

```
|————|————————————————————————|
| 1010,    | Extended entity data X, Y, and Z coordinates      |
| 1020,    |                                                   |
| 1030,    |                                                   |
|————|————————————————————————|
```

That's a valuable feature if you know when your "entities" are "extended" (I don't).

Naturally, extended entities in "world space" have their own kind of X coordinates:

```
|————|—————————————————————|
| 1011,   | Extended entity data X, Y, and Z coordinates of   |
| 1021,   | 3D world space position                           |
| 1031,   |                                                   |
|————|—————————————————————|
```

This will presumably be a handy feature if we ever find alien life on other worlds and need to keep track of which world we're referring to.

Adding up all of these codes, there are twenty different ways to write an X coordinate into a DXF file to describe a shape to be made. Woe unto a drawing program that uses the wrong kind of X when it generates a file.

And the woes of file format problems can be very real indeed. In 1999 the $125 million Mars Climate Orbiter was lost because of a file format incompatibility. The mission control was split between a spacecraft team in Colorado and a navigation team in California. During the first four months of the flight, file format errors kept them from exchanging files with commands for a key control system, so the teams e-mailed each other instructions. Once they fixed that problem, the computers could send command files to each other and be directly uploaded to the spacecraft. They considered this system to be more reliable than e-mail, because there was no human intervention. The problem was that the spacecraft and the California site were using metric units but the Colorado team sent English units, and that crucial bit of information was left unspecified in the file format. The spacecraft obediently followed these instructions and thereby flew so far off course that it was lost.

These file format problems reflect a casual assumption that is deeply mistaken: that computers compute, but machines only execute. These error-prone descriptions of things are not real languages that can contain instructions on how to interpret them, they're just cryptic lists that rely on advance agreement of their meaning. In "The Future" I'll look at an alternative approach, based on embedding intelligence into a tool. For today's tools that don't think, the intelligence must reside in the CAM software that controls them. To do this, CAM programs must embody the knowledge of how a skilled machinist works in order to properly direct a machine to do the same. This is generally as (un)successful as any other kind of artificial intelligence. CAM packages can appear to have been written by an IRS tax auditor on a bad day, requiring endless forms to be filled out:

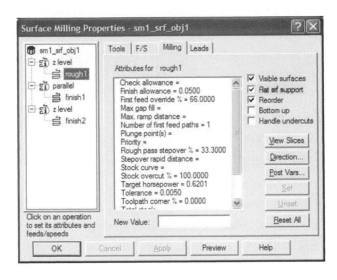

and then doing something apparently unrelated to what you expected.

The pathology of this state of affairs finally sunk in when I was in a rural Indian village setting up one of the early fab labs. We had sent the lab over with commercial engineering software, and were going through the laborious process there of installation, configuration, and training, when our host (Kalbag, described in "Making Sense") asked whether he could use that software in the next village, and in the village after that. I was about to contact the vendors to negotiate a license for a billion users in rural India, when it occurred to me that that approach might not scale so well.

The promise of personal fabrication for those billion users is in "open-source" hardware, meaning that they can collaborate on problem solving by sharing the files for a project such as the development of an improved agricultural implement or a healthcare sensor. But that won't be possible if it's necessary to issue engineering software licenses for a billion installations; beyond the cost, just the logistics would be prohibitive. The supporting software must be available to accompany the design descriptions.

Since there wasn't (yet) a developer community working on software for precision fabrication in rural villages, I started writing a CAM tool for use in the field, targeted at the needs of mortals, not machines. My first epiphany was, freed from the legacy of mainframes, just how easy this was to do. It's possible to do rapid prototyping of software for rapid-prototyping hardware, rather than retaining a mainframe mindset for the former. A second epiphany followed: instead of dedicated programs embodying assumptions about the applications of each kind of machine, a single CAM tool could talk to all of them. The same circuit, say, could be milled into a rigid circuit board, or cut by a sign cutter from a flexible substrate, or burned by a laser onto a conducting film. Here's what it looked like to use the CAM program, originally developed for Kalbag, to produce the hello-world toolpaths for the examples in the chapter "Subtraction":

sign cutter →   ← laser cutter

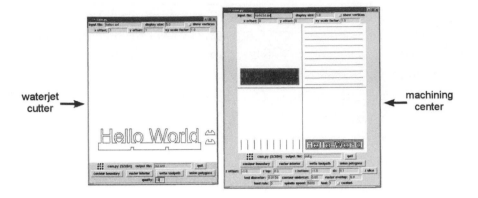

waterjet cutter →   ← machining center

The biggest surprise of all was that we found we liked what we were using in rural India more than the "real" packages back at MIT. Surely, I thought, the advanced fabrication needs on campus were far beyond the capabilities of any program I could throw together in a few months, requiring instead years of software development by teams of engineers. But, even more than in rural India, back at MIT I had been spending a growing fraction of my time arranging for the licensing, installation, and training for engineering software to connect each of the ways we design things with each of the ways we make them. Engraving pictures, wiring circuits, plotting with plastic, and writing with atomic beams were separate software worlds. Instead of acquiring a new program for each combination, we found that it was faster to add a new kind of input description or output process to a single shared CAM tool. The fab labs were teaching us how to work, rather than the expected converse.

There's a deep side to this story: a universal CAM tool is an example of a "phylotypic" bottleneck, a concept that originated in developmental biology. The idea is that there are many types of eggs, and many types of animals, but far fewer types of embryos within a group of related animals (a phylum). However it's done, the job of an egg is to assemble a relatively standard set of biological parts that can then be used to create a wide range of organisms, reducing the threshold for evolutionary improvements. The heart of this process is the Hox genes, master genetic switches that control what parts of the body end up where as the embryo develops. For any one organism there might be a more optimal way of growing, but repeated reinvention of development mechanisms would be detrimental to the viability of the species overall. Something similar happens in the Internet: there are many kinds of networks, and many kinds of applications, but they all pass through the IP protocol (described in "Communication"). Once again, for any one application there might be something better than IP, but devolving into a cacophony of incompatible standards would be collectively counterproductive. With IP, new networks and new applications can interface with, rather than reinvent, earlier solutions.

The evolution of a CAM tool from rural India back to MIT taught us that something similar can be true in fabrication: there are many ways to design things, and many ways to make them, but a short-term need to connect a variety of design tools with fabrication tools had the long-term consequence of freeing our engineering workflows from restrictive software assumptions about how a particular machine should be used. This made it possible to focus on learning how to use the machines together to solve problems, rather than on idiosyncratic ways to apply them separately. Engineering software can be an opportunity rather than an obstacle; when it becomes accessible to nonengineers it becomes possible in a fab lab, and a fab class, and a fab book, to teach almost anyone how to make almost anything.

The
Present

In the past, art became separated from artisans and mass manufacturing turned individuals from creators into consumers. In the future, there will be universal self-reproducing molecular fabricators. In the present, personal fabrication has already arrived.

This section is the heart of the book. Its twin goals are to introduce who is making what, and to explain how and why they're doing that. These stories are told through pairs of chapters introducing tools and illustrating their applications.

The tool tutorials progress through levels of description, from creating physical forms to programming logical functions to communicating and interacting with the world. Along the way, the associated applications range from activities in some of the poorest parts of the world (rural India) to the wealthiest (midtown Manhattan), from among the coldest (far above the Arctic Circle) to the hottest (a Ghanaian village), from addressing the most basic human needs (creating jobs) to the most refined (creating art), from whimsical inventions to ones that are needed for survival. These are all illustrated by introducing some of the pioneering people doing these things; each of them considers their emerging application for personal fabrication to be the most important one, and for each of them, it is.

The chapters in this section can be read as a business book, identifying emerging market trends. They can be read as a policy document, targeting where these technologies can have the greatest global impact. They can serve as a technical manual, reviewing currently available capabilities. They offer a portrait of a group of extraordinary people from around the world. And they provide a peek at the research frontiers for the enabling technologies. My hope is that they will be used for all of these purposes, providing enough background to follow up interest with action in each area.

This section really does represent the present. The people, projects, and processes portrayed here are a snapshot of what's possible today. They're a moving target; if the ongoing efforts in each of these areas are even remotely successful, the details will come to look very different. But what won't change is how people work, and how nature works. The fundamental human needs and machine mechanisms presented here are timeless and should serve as a relevant survey for a long time to come.

In these pairs of chapters, I review the use of tools that few readers will have access to. I've done that because this state of affairs is changing even faster than the technology. Increasingly, the greatest limitation on their widespread use is neither cost nor training nor research; it's simply the awareness of what's already possible. Hence this extended tour through the present, where the past meets the future.

Birds and Bikes

Irene

Irene Pepperberg is a pioneer in interspecies information technology. Irene has been able to show quite conclusively that Polly really does want a cracker when she asks for one. Irene found that African Grey parrots have the cognitive ability of roughly a five- or six-year-old child; they can learn hundreds of words; understand nouns, adjectives, and a limited number of verbs; and do abstract reasoning. This behavior matches the abilities of great apes and dolphins, and it is beyond that of monkeys (and even some people I know).

A student working with Irene, Ben Resner, took my course "How To Make (almost) Anything." For his semester project he developed a computer interface for parrots, which he called the Interpet Explorer. He was motivated by the observation that parrots are social animals and can quite literally go crazy when left home alone. Ben, consulting extensively with the parrots, built a computer interface that could be controlled by a beak (I hesitate to call it a mouse). This had tabs that the birds could push or pull, controlling a program that allowed them to choose from among their favorite activities. The birds could use this

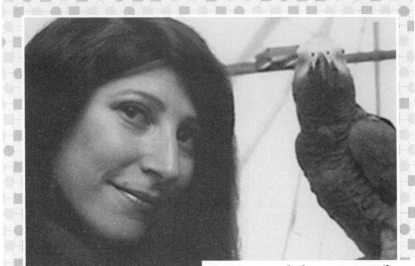

Irene and Alex, a star pupil

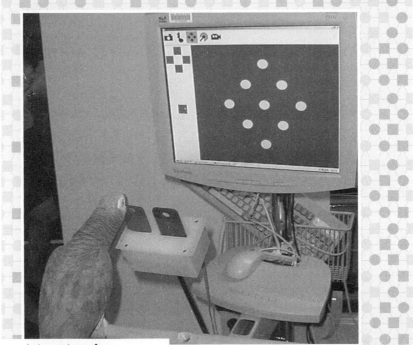

Interpet explorer

system to see Irene on video, or see other animals, or listen to music, or even use a kind of drawing program written for them.

Ben used a laser cutter to make the parts for the Interpet Explorer. A laser cutter is a bit like a laser printer, but the laser is so powerful that instead of just writing an image it can engrave the image into a material, or cut right through it. It can do that so accurately that the parts it makes snap together, allowing quick assembly of three-dimensional structures such as the Interpet Explorer from sheets of two-dimensional materials. This allowed Ben to test many candidate schemes for parrot controllers until he settled on one that worked for the parrots.

The parrots loved it. After all, as social and intelligent animals they have very real and often unmet communication needs. As a population, they've been underserved by information technologies. Our corporate research sponsors at MIT, on the other hand, thought that this sounded awfully frivolous. That is, until they thought about their existing markets. Manufacturers of cell phones and laptops have already sold their products to every two-legged creature that they can find. These companies are killing themselves trying to persuade their existing customers to pay for incremental upgrades to new products, while fighting with other manufacturers over new customers. All this gets measured in ebbs and flows of tiny slices of market share. The pet market, on the other hand, is worth billions of dollars a year, and it's currently not possible to buy much more than food dishes and furry toys. What pet owner wouldn't want to be able to call in to say hello to their pet over lunch? What homebound pet wouldn't want to be able to use technology in the home as their owners do, to play games, watch videos, and communicate with friends? Like Kelly's ScreamBody, Ben and Irene's personal fabrications are likely to interest many more people (and animals) than impersonal products on the market today.

Saul

Saul Griffith came to MIT as a grad student from Australia. Unlike most grad students, though, Saul has a penchant for making himself part of his experiments. While others fly kites from the ground, Saul likes to fly kites that can lift more than his weight and pull him into the air along with the kite. Naturally, he returns to the ground when the kite does, which can happen a bit faster than planned since he's trying to fly the kite while flying himself.

Saul's taste for extreme activities extends to the tools he uses. He quickly became a passionate user of our waterjet cutter, which is a bit like a laser cutter on steroids. Instead of using light to etch or cut, it shoots out a supersonic jet of water containing tiny abrasive particles. These are going so fast they can blast through materials that lasers can't get through, like thick metal plates, glass, or stone.

Saul of course wanted to make something on the waterjet that could carry him, so he set out to develop a process to "print" a bicycle. Like a laser cutter, a waterjet cutter functions much like a printer, but with the ability to engrave and cut parts as well as draw images. And, like a printer, a waterjet cutter is as fast at making many different things as it is at making one thing over and over again. Saul calls the laser and waterjet cutters "tools of mass construction," and he is passionately pursuing the implications of access to them around the world.

As Ben did with the parrots, Saul quickly tested many different bicycle designs, using himself as the test subject. He settled on a process using a tough, clear polycarbonate plastic to assemble the bicycle frames, and then attaching conventional bicycle components to them.

Saul's experiments turned into a student-run and -taught course, in which each participant designed and built a custom bicycle. Since they were each starting from a clean sheet (of polycarbonate) rather than just cutting the tubes that are used to make conventional bicycles, the students quickly started personalizing the form as well as function of

Building a bike

Personal transportation

their bicycles. They came up with wonderfully expressive designs; my favorite was a rideable version of Monet's reclining nude. These bikes began to appear around Cambridge as a fashion accessory as well as a means of transportation. Their riders' mobility was limited only by the inevitable questions they got about where one could buy such a cool bicycle.

Alan

Among the people who got hooked on rapid prototyping after seeing one of these bicycles was Alan Alda. He's a serious student of emerging technologies and comes by MIT periodically to see what's new. During an interview about personal fabrication, we showed him one of Saul's bikes as an example. As soon as he saw it he was on the bike and we watched him recede into the distance for a test spin. He came back raving, having realized that if he had a waterjet cutter at home he could receive the bicycle design by e-mail and produce one on his own machine. Saul had in fact done exactly that with his sister in Sydney.

A flash periscope

The interview immediately ended when Alan realized that instead of just talking about personal fabrication we could be in the lab actually doing it. He wanted to test an idea he'd had for an invention but never been able to make: a flash periscope.

A test picture without the flash periscope

A test picture (of a proud inventor) with the flash periscope

The demonic red eyes that are the bane of flash photography are caused by light from the flash being reflected back into the camera from the eyes. Alan reasoned that if he could just elevate the flash, its direct reflection would aim away from the camera. That was easy in the old days when the flash was separate from the camera, but how to do it for an integrated flash? Alan proposed to make a custom periscope for the flash.

We were soon seated at a laser cutter, sketching designs and turning them into kits of parts to assemble a periscope that fit onto his camera. After taking a test picture of the red eye produced by the unmodified camera, we attached the periscope, took another picture, and lo and behold . . . it didn't work. On closer inspection of the picture we saw that the camera was aiming the flash too low. Back at the laser cutter we changed the design to have an adjustable mirror angle, and this time it worked beautifully.

Grace and Eli

The red-eyed subject of the preceding picture is one of my two favorite shop users, my son, Eli. Since he was five, he and his twin sister, Grace, started coming to MIT with me to make things. They were in the lab with Alan Alda to show him some of their projects.

In the shop, Grace and Eli would sketch designs they wanted to make, and my job was to transcribe the designs for the appropriate fab tools. Their favorite is the laser cutter because of its speed in satisfying eager but impatient young designers. Eli's first project was for his favorite stuffed animal, a bear named Bear Bear. Eli decided that Bear Bear needed a chair to sit in. And Grace's beloved doll, Cherry, had everything a doll needs but for a cradle. Creating both of these objects was an afternoon's activity, using the laser cutter to assemble prototypes from parts, and fine-tuning the designs to satisfy Eli, Grace, Bear Bear, and Cherry.

Over time, these projects have led to a great phrase that the kids use: "go to MIT." They say this when they think of something to build: when Eli wants a pirate ship that can sink on cue, or Grace wants to add some parts to a play set, the solution is to "go to MIT." Instead of limiting themselves to what can be found on the shelves of a toy store, they're limited only by their active imaginations.

While their inventions do sometimes stray a bit beyond what's feasible (we spent a while discussing possible approaches to making a horse) or even physically possible (we haven't made much progress on creating

Bear Bear's chair

Cherry's cradle

a room without gravity), they're growing up thinking of engineering as something that they do for themselves rather than expecting others to take care of it for them. And in believing that, they're teaching me about the profound opportunity and demand for tools of mass construction for all of the people on the planet who can't go to MIT.

Subtraction

Computer-controlled cutting tools provide a quick (although not necessarily cheap) thrill in physical fabrication. As with cooking a dinner or managing a business, it's usually far faster and easier to throw out what's not needed than to build something up from scratch. This first of the tool chapters will introduce subtractive cutting machines, followed in the next one on additive printing processes.

Start with one of the key tools in any well-equipped nursery school, a pair of scissors. The computerized equivalent, called a sign cutter, puts a knife blade onto a moveable arm. As a day job these machines make lettering for signs, but like their nursery school counterparts, they can cut many materials other than the vinyl that's commonly used for signs. For example, it's possible to cut the words "Hello World" into a copper sheet:

Once cut, the letters can be lifted off of their paper backing with a sticky transfer tape:

and applied to another material:

The most common material used in sign cutters is vinyl for graphics, but plotting in copper is a quick way to make functional flexible circuits, antennas, and wiring for interconnections. Or, by using a stiff card stock, a pop-up folding "Hello World" could be cut out.

Sign cutters are relatively cheap and easy to use, but they are limited to materials that a knife can cut through. Beyond that it's possible to spend nearly arbitrary amounts of money on subtractive tools that cut with high-powered lasers, or jets of water, or rotating tools, or beams of atoms, or jets of gas, or most any other way anyone has ever found to get rid of anything. Of these, the one that arouses the most passion in its users is the star of the last chapter, the laser cutter.

A laser is an amplifier for light. It takes a little bit of light and turns it into an intense beam, which can be focused down to a tiny spot. By putting so much energy into such a tiny space, it can burn, melt, evaporate, or ablate away the material it encounters. And because the "tool" is a beam of light, it can move very quickly, and make cuts as narrow as the focus of the beam.

The most common kind of laser for cutting applications uses carbon dioxide as the amplifying medium. A radio signal travels through a tube containing the gas, exciting it so that a single photon can stimulate a molecule to emit a second photon, and each of those can stimulate the emission of still more photons, in a cascade that efficiently converts the energy in the radio signal into a concentrated beam of light. With carbon dioxide, the color of this light is below the range that the eye can see, so the beam is visible only by its impact on whatever it hits.

A typical laser in a laser cutter will produce about a hundred watts of light. This is *much* more than a one-hundred-watt lightbulb, because most of the power in a lightbulb goes into heat rather than light, and the light that is produced goes off in all directions rather than being concentrated into one direction. One hundred watts of laser light is enough energy to cut through plastics such as acrylic and soft materials such as wood and cardboard. It's not enough for harder materials such as metal or stone; cutting them requires either much more powerful and expensive industrial lasers or other kinds of cutting tools, like the waterjet cutter to be covered next. Also, laser cutters are not safe to use with materials that emit dangerous fumes when they burn, such as the PVC plastic commonly used in pipes, which will emit chlorine gas.

At low power, laser cutters can mark or process materials. When the laser cutter was first installed in Boston's South End Technology Center (described in the next chapter), one of the first hello-worlds done there used it as a state-of-the-art computer-controlled personalized toaster, selectively heating a slice of bread:

At higher powers, laser cutters become fabrication rather than decoration tools by cutting all the way through the material. Here's one cutting a piece of acrylic, viewed through the safety glass that encloses the laser. The only clue to the energy being invisibly aimed at the sample is the bright spot where it's burning away the plastic:

The laser is cutting out a hello-world again, this time with a base and feet:

These snap together to make a stand:

Standing the hello-world upright illustrates the real power of laser cutters. In the few seconds it took to make the notches and cut out the feet, the laser cutter transformed the two-dimensional sheet of acrylic into a three-dimensional object.

The laser spot size used to cut out this hello-world was just 0.010 inch (10 thousandths of an inch, or 10 mils). That spot pulses thousands of times a second, each pulse burning away a little bit of material, making a continuous cut as the mirror aiming the laser moves. The mirror is positioned to better than 0.001 inch, 1 mil. That's the tolerance needed to make a structure that fits together without needing fasteners or adhesives. If the gap between the parts is any larger than that, they won't stay together, and if it's smaller, then they won't fit together at all. This kind of press-fit assembly is increasingly important in industrial practice for making everything from toys to airplanes, because it's faster and cheaper than using screws or glue, plus it's easier to disassemble things made that way.

Laser cutters can also produce less welcome results. They direct an awful lot of energy into a small space, which can ignite not just the spot but everything else around it, up to and including the building. When everything is working properly, the airflow in the laser cutter will prevent the material being cut from being ignited. But if too much power is used with a flammable material with inadequate ventilation, the results can be dramatic:

This fire occurred when a student at MIT started cutting an object and then left the laser cutter unattended to go off to his office, never a particularly good idea with a dangerous machine. He realized that he had used too much laser power for the material that he was cutting when he looked out the window and saw the fire engines arriving.

The force of the spray from the fire hose that put out that fire can itself be used for fabrication. A waterjet cutter picks up where the laser cutter leaves off, sending the flow from what is essentially the pump of a fire engine through a nozzle that is tens of mils, rather than inches, across. Sending that much water through an opening that small produces a supersonic stream. On its way out, tiny abrasive particles mix with this jet, typically garnet a few mils in size. These particles do the cutting. When they accelerate to the speed of the jet, they pick up so much energy that they blast through just about anything they encounter. Steel, glass, or ceramic—it doesn't matter. The waterjet can cut through thick, hard, brittle materials that defeat lesser tools. And the combination of the small orifice with the high speed of the particles results in a cutting tool that acts much like the beam of a laser cutter, capable of making fine cuts with tight tolerances for complex shapes.

About the only thing unimpressive about a waterjet cutter is seeing one in action. Here's the business end of a waterjet cutter ready to cut an aluminum plate:

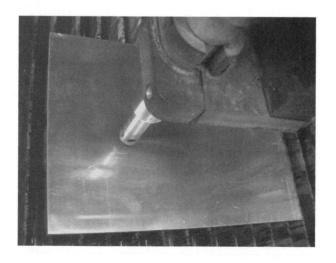

The high-pressure plumbing contains a 14-mil orifice that is made of sapphire to stand up to the forces on it. The cutting itself is done underwater, to contain what would otherwise be a very impressive splash. The tank fills with water before the jet is turned on, and a foam collar around the nozzle helps further contain the splash. The only sign of the jet's cutting the submerged plate are the unremarkable bubbles it produces:

Using the same file that the laser cutter used, the result is the same "Hello World," but it's now made of aluminum:

With the waterjet cutter, the material could just as easily be glass, or stone, or most anything else. Because of this enormous flexibility, waterjet cutters are taking over work that's traditionally been done by other machine tools, particularly the numerically controlled (NC) milling machines that date all the way back to the pioneering experiments interfacing the original Whirlwind computer at MIT.

NC mills have been the industrial workhorse for subtractive fabrication. They rotate a cutting tool called an end mill, which is much like a drill bit but can move horizontally as well as vertically. If the tool can move left-right, front-back, and up-down, it's called a three-axis mill; a four- or five-axis mill adds rotations of the material being cut.

The great advantage, and disadvantage, of an NC mill over a water-jet or laser cutter is the presence of a cutting tool. The advantage of the NC mill is that the tool has an end, allowing it to cut to a carefully controlled depth. Unlike a beam of light or jet of water, this tool can contour three-dimensional shapes to precise dimensions, with a gorgeous surface finish. And the NC mill's disadvantage is that if the tool is dull, or is moving too fast or too slow, or isn't properly lubricated, it will cut poorly. Or possibly not at all: If an end mill is pushed too far beyond its limits, the stock can cut the tool instead, shattering the brittle steel.

There are nearly as many kinds of end mills as there are applications of them. Here are two common designs:

The first is a flat-end mill for making flat surfaces. The flutes meeting in the middle of the bottom face allow it to cut downwards like a drill, but unlike a drill, this end mill leaves a flat bottom. It can also cut with the flutes along its sides to make flat walls. The second design is a ball-end mill with a rounded bottom, used for making curved surfaces, smoothing the steps produced by succeeding passes of the tool as it follows the contours of the surface. End mills can range in diameter from inches to thousandths of an inch, turning at speeds ranging from thousands to hundreds of thousands of revolutions per minute.

The diversity of end mills is matched by the range of available milling machines. Here's an inexpensive tabletop mill carving "Hello World" out of a dense foam:

Easily machined foams and waxes are frequently used to prototype products, or to make molds for casting other materials.

Here's its bigger sibling, an automobile-sized (and automobile-priced) machining center, making an aluminum hello-world:

Larger mills like this not only can make larger parts; they can add automatic tool changers to switch between multiple tools for complex jobs.

The biggest NC mills of all are the size (and price) of a house, where the operator actually travels in a cab next to the tool:

Perfect for personalizing a jumbo jet airframe.

The unsung hero of all of these cutting tools is their metrology, the measurement devices that let them monitor what they are actually doing. A twenty-dollar CD player positions the pickup head to within a millionth of a meter, a micron. Any self-respecting machine tool, even a house-sized one, now uses position sensors with comparable performance, placing the tool with a resolution of a few tenths of a mil. A hair is huge compared to this, a few mils thick, or tens of microns. Such micron-scale control is what allows all of these machines to produce parts with such exquisite tolerance and finishes. The accessibility of micron-scale metrology is what makes fab labs possible, from the press-fit three-dimensional structures made on the laser cutter to the fine features of surface-mount circuit boards machined by a humble tabletop computer-controlled mill.

Growing
Inventors

There's a demand for personal fabrication tools coming from community leaders around the world, who are embracing emerging technology to help with the growth of not only the food and the businesses in their communities but also the people. This combination of need and opportunity is leading them to become technological protagonists rather than just spectators.

Mel

Mel King is a grand not-so-old man of Boston community activism. He helped invent modern urban planning quite literally by the seat of his pants.

In 1968 Mel led the "Tent City" encampment. The city of Boston had conveniently sold a central plot of land to a former fire commissioner, who tore down the low-income housing that was on the site to make room for a huge parking lot. Naturally, the former residents

Community
center

didn't think that this was such a good idea. To protest, they occupied the site, and as the protest dragged on they erected tents. This encampment came to be known as Tent City. Their persistence forced the inclusion of low-income housing in the complex, which was eventually named Tent City in honor of the protest that created it. The example of Tent City helped establish subsequent zoning requirements for mixed-income housing development on city, state, and federal land.

Mel now runs the South End Technology Center at Tent City (SETC). He's made a career of doing for technology what he did for urban low-income housing, helping make it accessible to everyone. SETC operates as a community center, open day and night to anyone interested. People can come in off the street to just hang out, or sign up for classes when they want to learn more to satisfy a personal interest or acquire a marketable skill. Classes are free except for the cost of materials and certifying exams.

Mel had found that the TV networks weren't telling the stories of people like him or his neighbors, so he started working with community organizers and city councilors to provide cable access for programs produced locally "by the neighbors and for the neighbors."

He moved on to providing computers and classes at SETC when he saw that the digital revolution threatened to bypass his community. From there he added an activity to make computers available to individuals and groups within and beyond his community by assembling working computers out of parts from discarded ones.

A natural next step was to set up a fab lab at SETC, aimed at democratizing not just the use of technology but also its development. As the machines arrived, hands-on training began with hello-world examples (of course), cutting and printing text with the tools. These were intended to be no more than exercises before the serious work started, but there turned out to be much more interest in them in their own right. The ability to make a tangible hello-world represented a real opportunity to personalize fabrication.

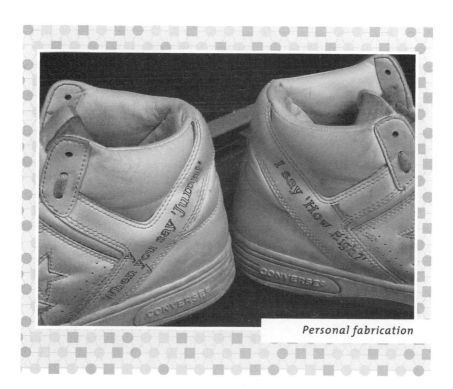

Personal fabrication

People, particularly kids, at SETC put personally meaningful words and images on anything they could get their hands on: jewelry, shoes, tools, and more. The lab began producing custom pop-up greeting cards, and dioramas, and "Adinkra" symbols sent from the Ghana fab lab. All of this activity was like a chorus of technological voices saying "I'm me. I exist."

After going through the initial assortment of high-tech materials supplied with the cutting tools, Mel wondered how SETC could keep providing the materials to feed the machines. The solution to this problem turned out to be an opportunity waiting on their doorstep: trash. Or, rather, all of the high-performance materials that were thrown out or left for recycling: cardboard boxes, plastic packaging, plywood containers, and so forth. These are engineered to maximize some combination of weight, strength, surface finish, and cost, and like any city, Boston is awash in them.

Community access to the precision fabrication tools in Mel's fab lab turned shipping boxes from trash into a valuable raw material that could be remanufactured into something else, like the high-tech crafts the girls at SETC sold, with implications ranging from personal empowerment to communal recycling to urban economic opportunity. These are big things in communities that have little hope. As Mel says, "The rear wheels of the train don't catch up to the front wheels of the train unless something happens to the train." At SETC, personal fabrication offers an opportunity to speed up the back of the train.

Nana Kyei

Nana Okotwaasuo Kantamanto Oworae Agyekum III is an Omanhene, a Ghanaian paramount chief, known in private life as Kyei Amponsah (pronounced "chay am pon sa"). "Nana" is an honorific title for a chief; below him are the divisional chiefs and lesser subchiefs. Being a chief in

Nana

Ghana is not just a job; it's a way of life and a lifetime responsibility. The chief is his ethnic group's spiritual, political, and judicial leader, a revered custodian of their cultural heritage, and a living spiritual and genetic link to his ancestors.

In the south of Ghana the chiefs are selected by matrilineal succession. When a new chief is needed the queen mother, consulting with the town's "king makers" (elders), selects the most promising member of the royal family and "captures" him. Being selected is a great honor but also a mixed blessing, as it entails a commitment to living a very different kind of life.

The installment of a new chief is referred to as an "enstoolment." The chief's seat of power is represented by a stool rather than a throne, which is much more convenient if you live in a rural village rather than in Buckingham Palace. Along with the stool comes a host of trappings of authority that follow the chief around. One person carries an umbrella so that the sun can't directly fall on the chief. Another, a "linguist," talks to and for the chief, since commoners aren't allowed to converse directly with the chief; even if they're standing next to each other, the linguist must relay the conversation.

When I first met Nana Kyei in Ghana he was traveling incognito, dressed as an ordinary person so as not to be encumbered by these formal trappings of office. He grew up in Ghana and Great Britain and then moved to the United States, where he studied and subsequently worked in the IT industry. Kyei was employed as a high-tech consultant when he was selected to become a chief. He was summoned back to Ghana, where his prior and future careers merged in his desire to bring the chief's role from the nineteenth to the twenty-first century.

Around the turn of the twentieth century, the authority of the chief was gradually usurped by the occupying British colonial administration. It was convenient to retain the chiefs in order to preserve the appearance of continuity, and they were valuable as a channel to convey colonial instructions implementing Britain's indirect rule. This arrangement was formalized in their incorporation into a Native Administration. But the British took over any function that could threaten their political and economic control, most important, the valuable coastal slave trade. The chiefs were left with authority over traditional customs, rituals, and festivals. Even worse, as part of a "divide and rule" strategy appropriate for absentee landlords, rivalries among the tribes helped fragment the country and keep the British in power.

By the time of Ghana's independence in 1957, the typically rural chiefs were shunned by the emerging British-educated urban elite. People, political power, and economic activity migrated to the cities, leading to a decline in Ghana's essential agricultural productivity.

However, the position of the chief was still revered within their communities. When I met Nana Kyei in 2004, he had a dream that that status could become the foundation for a new authority, based not on politics or money, but rather on ideas. He hoped to lead his people by marrying emerging technology with local needs, much like the work he did as a consultant.

When a fab lab arrived in Ghana, based at the Takoradi Technical Institute, Nana Kyei appeared soon after. He had cleared his schedule to come and discuss possible projects to help address those local needs. The two greatest problems he presented were providing access to information and energy. Where Internet access was available, a 64-kilobit-per-second connection (equivalent to a fast dial-up line) cost the equivalent of $450 per month. In most places it wasn't even available. And the energy situation was if anything even worse.

Most of Ghana's electricity comes from hydroelectric power generated from Lake Volta. In 1996 about one gigawatt was produced for the whole country, equivalent to a single large power plant in the United States. This supplied just 10 percent of the country's energy needs; the bulk, 70 percent, came from firewood. That firewood is provided by chopping down forests, an unsustainable practice that drives up the price and drives down the quality of the wood, while incurring all of the costs of deforestation.

In Ghana I visited a small local distillery powered by wood. A load of wood is delivered to the distillery every two to three weeks, when it can

Power plant

be found, and costs 500,000 cedis, or about fifty dollars. That corresponds to an annual energy expenditure of a few thousand dollars for the distillery, in a country with a per capita income of a few hundred dollars.

Meanwhile, a chief's umbrella protects him from Ghana's intense sun, which at its peak provides a thousand watts of power per square meter. Averaging the solar radiation over day and night, clear and cloudy weather, it's still a few hundred watts. My own house consumes an average of about a kilowatt of electricity, which could be provided to a house in Ghana by collecting the sunlight falling on just a few square meters.

The use of solar energy in Ghana today falls into two general categories. There are simple concentrators that focus sunlight for cooking but aren't good for much more than that. And there are photovoltaic panels that produce electricity at a cost of a few hundred dollars for a few hundred watts. Since the price of a photovoltaic panel equals the average annual income, there aren't too many takers.

As Nana Kyei described the demand and available solutions for solar energy, it was clear that neither of the existing approaches could quickly scale to meet the national energy need. The cooker does too little: all it can do is cook. And the photovoltaic panel does too much. The electricity it produces is expensive to generate and to store, and then requires the use of expensive appliances. Furthermore, energy is lost at each of those steps so that the overall efficiency from collecting to using the solar power is low.

However, sunlight can do many things other than cook food and make electricity. In between the simplicity of making a solar cooker and the complexity of producing semiconductor photovoltaic cells, the dimensional control of precision fabrication tools offers new ways to harness the sun's energy, breathing new life into some old ideas. One, familiar to any kid who has played with a magnifying glass on a sunny day, is to use focused sunlight as a cutting tool. If a computer rather than a hand directs the focus, sunlight can be used as a freely available alternative to the expensive and power-hungry industrial lasers found in laser cutters.

Another idea dates back to 1913, when the eccentric inventor Nikola Tesla patented a new kind of turbine. Conventional turbines work like airplane wings, generating a force by redirecting the flow of air or water. Producing the sets of fixed and moving blades in a turbine is a challenging task. Tesla's elegant design uses a much simpler approach.

When a fluid flows past a surface, there's a transition region called the *boundary layer*. Right at the surface, collisions between the molecules in the liquid and the surface slow down the molecules so that the flowing liquid is effectively stopped at the surface. A bit farther away from the surface the molecules are slowed by

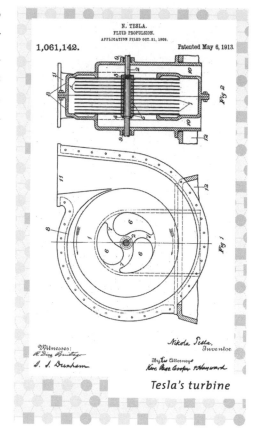

Tesla's turbine

those near the surface, and those in turn slow down others that are farther away, creating a velocity gradient between the freely flowing liquid and the fixed surface. The gradient in the velocity exerts a drag force on the surface.

The boundary layer thickness is small, less than a millimeter. Tesla's design is based on disks spaced by roughly that width. He realized that if the plates were much farther apart, then the liquid could flow easily between them, and that if they were much closer than that, then the liquid could not flow through at all. But if they were spaced by the width of the boundary layer, then the attachment of the boundary layer to the disks would result in an efficient transfer of energy from the liquid to the disks.

These "Tesla turbines" are not commonly used today because boundary layer attachment breaks down at high flow rates; it wouldn't work in a jumbo jet engine, for instance. But at slower flow rates it does work well, using precisely spaced disks to exchange energy between a liquid or gas and the turbine.

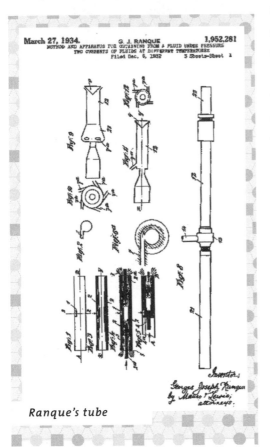

March 27, 1934.   G. J. RANQUE   1,952,281
METHOD AND APPARATUS FOR OBTAINING FROM A FLUID UNDER PRESSURE
TWO CURRENTS OF FLUIDS AT DIFFERENT TEMPERATURES
Filed Dec. 6, 1932   3 Sheets-Sheet 1

Ranque's tube

If instead of boiling water to cook with, a parabolic dish concentrated sunlight to boil water to produce steam to power a turbine, then the rotary motion would be available for all kinds of mechanical work. It could directly run machinery, pump water or lift weights to store energy, power an electrical generator, or turn an air conditioner's compressor.

Nana Kyei underscored the importance of cooling in a hot country, for food and medicine as well as people. Roughly a third of Ghana's food spoils in transit for want of refrigeration. Solar heating is easy, but what about solar cooling? That's also possible, thanks to another old idea with new applications enabled by the tools in a fab lab. The vortex tube was accidentally discovered in 1928 by George Ranque, a French physics student. In a vortex tube, compressed air is blown in transversely much as a flute player does, but in this tube a precise spiral shape sets it spinning as fast as 1,000,000 RPM. This motion separates hot air toward the outside of the tube from cold air near the axis. Exhausts can then bleed these off separately, providing streams of hot and cold air.

The vortex tube effectively sorts the fast (hot) and slow (cold) molecules in the airflow. In industrial processes, vortex tubes generate temperatures down to -50° Fahrenheit. They're not used for air conditioning because they're less efficient than a conventional cooling system based on compressing a refrigerant; but unlike a conventional air conditioner, a vortex tube has no moving parts and no chemicals, so they're handy if compressed air is already available.

The possibility of making vortex tubes with fab lab tools led me to bring one to a "Kokompe" (garage district) in Ghana. Here, workshops build and repair the trucks that carry the food that's spoiling in transit. When the tube was connected to their compressed-air system, it nearly caused a riot. Workers came running to feel the cold air coming out of what appeared to be a simple tube. There was no discussion about the relevance of a fab lab for them; their questions leapt to the capacity and efficiency of a vortex tube, whether it could be driven by a truck's exhaust air, and how quickly they could start making such things.

Vortex tubes and Tesla turbines are at the fringes of mainstream engineering, mixing quirky inventors, shady businesses, and murky

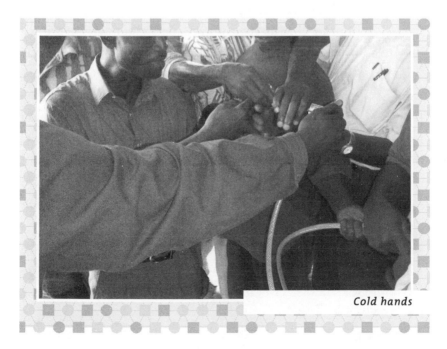

Cold hands

explanations with serious science and even more serious applications. From a Ghanain village to an MIT laboratory, Nana Kyei's interest in emerging energy technology helped launch ongoing exploration of how these kinds of precise structures can help harness the sun's energy.

The back-of-the-envelope numbers are staggering. Collecting a conservative 100 watts per square meter, Ghana's land area of 238,533 square kilometers corresponds to 10^{13} watts of available power, or the output of ten thousand nuclear reactors. If computer control over the fabrication of complex shapes can produce collectors that capture even a tiny fraction of this, Ghana would go from being a poor country to one of the richest in one of the most important natural resources, energy. Powering an economy with light rather than oil would have consequences far beyond Ghana's borders, in the geopolitical implications of energy economics. Power truly can flow from a chief's ideas.

Anil

Anil Gupta is a global guru of grassroots invention. From his base at the Indian Institute of Management in Ahmedabad, he's made a career out of finding and assisting the inventors hidden in plain sight.

Anil runs the "Honeybee Network." Modeled on how honeybees work—collecting pollen without harming the flowers and connecting flowers by sharing the pollen—the Honeybee Network collects and helps develop ideas from grassroots inventors, sharing rather than taking their ideas. At last count they had a database of ten thousand inventions.

One Indian inventor couldn't afford or justify buying a large tractor for his small farm; it cost the equivalent of $2,500. But he could afford a motorcycle for about $800. So he came up with a $400 kit to convert a motorcycle into a three-wheeled tractor (removable of course, so that it's still useful as transportation). Another agricultural inventor was faced with a similar problem in applying fertilizer; his solution was to modify a bicycle.

Invention guru

Farm transportation

Power from the front and rear of a cow

One more inventor was struck by the absence of electricity and the presence of bullocks in rural India. These ubiquitous male cattle are the engine for the agricultural economy, pulling its implements and hauling its loads. He wondered why their power wasn't harnessed to generate electricity.

His solution was a contraption that connected a bullock to a generator and converted its walking motion around a track into electricity. One of these bullock-powered generators could produce about a kilowatt of electrical power, plenty for a farm or a few houses. And the bullocks didn't seem to mind doing this; it was just one more thing for them to pull.

As an energy source, there's a lot to like about cattle compared to coal. Cattle are a sustainable resource (as long as they're fed). And they're a renewable resource—no other kind of generating machinery is capable of reproducing itself. Even their waste is useful: beyond fertilizer, this inventor found that he could use dung as an electrolyte to make batteries.

It was a humbling experience when Anil first introduced me to a group of these grassroots inventors. I rather naively thought that I would bestow a few nuggets of technical wisdom on them. I introduced the kind of personal fabrication research being done at MIT and in the field and then invited questions. One gentleman mentioned that he was having trouble with a grain impeller that he made and wondered if I had any suggestions. This machine had a long screw to feed the grain, and the screw was quickly wearing out under the force of pushing the grain.

I told him about an advanced technique for strengthening the surface of metals, based on growing a ceramic coating. I was disappointed when he looked at me blankly; I feared that I had confused him with a high-tech approach that was locally inappropriate. But after a moment he replied that *of course* the right thing to do was to grow a ceramic; he was looking for guidance on *which* ceramic. Similar questions from his peers soon followed on self-lubricating alloys, and materials that can maintain a constant temperature by changing phases. On their own they had managed to pick up a substantial materials science background, and were asking questions that went well beyond my own knowledge.

These inventors didn't need to be told about the possibility of personal fabrication; they already made things for a living. The opportunity that they saw for connecting computation with fabrication lay in improving dimensional control. They could conceive of things like motorcycle tractors and bicycle sprayers but had trouble making the reality match their imagination. The parts that they cut and joined never came out quite where they planned. Beyond any one invention or technique, they wanted and needed the modern industrial practice of "net shape" manufacturing. This puts materials exactly where you do want them, rather than removing them from where you don't. That's the domain of additive fabrication processes, introduced in the next chapter.

Mel King, Nana Kyei Amponsah, and Anil Gupta have almost identical roles in their very different worlds. They don't work as hands-on inventors, but they share an unerring sense of technological possibilities that's better than many of those who do. Each of them leapt at the opportunity for early access to personal fabrication tools in their communities, seeing in this technology not a discontinuous advance but rather a natural next step in their lifetime journeys. Under the capable guidance from them and their peers around the world, these tools can help develop the planet's most precious natural resource of all, its people and their ideas.

Addition

Additive fabrication technologies are to subtractive technologies as potters are to sculptors. Instead of chipping away at a stone to produce a sculpture and a studio full of stone chips, potters add just the clay they want, where they want it. This old idea is at the heart of some of the newest manufacturing technologies.

Subtractive cutting tools necessarily produce waste as well as parts. It's possible to effectively run those processes in reverse, building a part up from its raw materials by selectively injecting, or ejecting, or extruding, or depositing, or fusing, or sintering, or molding them. These additive approaches all start with nothing and end up with something, ranging from simple packaging prototypes to complete working systems.

One of the simplest and most common approaches to building without waste is neither additive nor subtractive; it could be called *equality fabrication*. The idea is to deform one part around another one, ending up with the same amount of material in a new shape. One such technique is vacuum forming, which heats a sheet of plastic to soften it, and then uses a vacuum to pull the sheet around a desired shape. Another technique, blow molding, uses compressed air to push the softened material against the inside of a mold; that's how bottles are made.

Here's a vacuum former, with a plastic sheet ready to be molded around the raised hello-world letters machined in "Subtraction":

The clear plastic sheet is suspended above the form. After a lid is brought down to heat the plastic, and the air underneath the sheet is evacuated through the holes in the base, the sheet will wrap around the text:

When the sheet separates from the mold, an impression of the letters has been transferred to it:

Because vacuum forming and blow molding are so fast and cheap, they're the techniques of choice for producing the packaging that contributes so much to our landfills.

If the plastic is heated still further and compressed even harder, it can flow like a liquid into a mold. This is done in injection molding to define a solid object rather than just a surface, and can be used to mass-produce the most intricate shapes. The molds are usually made out of mating halves that can be separated to remove a part after it's been injected into the mold. Here's an injection mold that uses the relief hello-world text machined in "Subtraction" as one of the sides:

The cavity in the other side of the mold will form the base for the letters. Molten plastic is forced in through the external opening in the mold, called the *sprue*. It then spreads through a channel called the *runner* and into the cavity through *gates*. Because the plastic that's initially injected isn't completely melted, the runner overshoots the gates so that the mold fills with plastic that is flowing smoothly.

The injection molder tightly squeezes the halves of the mold together so that plastic doesn't leak out from the seam between them. Plastic pellets from a hopper are melted and forced into the mold by a heated screw:

Once the plastic has cooled, separating the sides reveals the molded plastic now contained within:

The plastic that remained in the runner can be snapped off at the gate (for later reuse), leaving the injection-molded hello-world made from the plastic that filled the cavity and flowed into the machined letters:

Now, though, instead of making just one of these parts, it's as easy with the injection molder to produce a whole chorus of hello-worlds. It can take hours or even days to make a mold, but only seconds to shoot the plastic into it, so the extra effort is quickly justified for any kind of production run.

The challenge in injection molding is to obtain the best possible surface finish with the least possible postprocessing or wasted material, and the fewest apparent imperfections from the mold's seams and connections. At a company that lives on injection molding, like LEGO, they treat the mold makers like high priests. They work in an exclusive inner sanctum, making molds with the finish of a gem and the tolerances of a telescope. Done right, an unblemished perfect part pops out of the mold needing no further finishing. Done by lesser mold makers, the result is a mass-produced mountain of mediocre merchandise.

Many things other than just plastic can be injected into a mold in order to modify the physical properties of the part that's being made. Hollow spheres added to the plastic can reduce its density and hence its weight, or fibers mixed in can add strength because it's hard to stretch them. Components that conduct electricity can be molded by mixing in electrically conducting additives. It's even possible to use injection molding to make high-temperature ceramic parts or strong metal parts. A ceramic or metal powder, mixed with a plastic binder to help it flow, is injected to produce what's called a "green" part. This can then be heated and compressed (sintered) to burn off the unwanted plastic and fuse the particles together, leaving behind a solid part. Or, more important, lots of parts, since molding is far faster than machining.

If, on the other hand, the goal is to make one thing rather than one million, then mold making is a lot of extra work. Countless rapid-prototyping techniques have been developed to do just that: quickly make a prototype of a part to check it before going to the trouble of making a mold. This is also called *3D printing*, because rapid-prototyping machines are used like printers but make three-dimensional objects rather than two-dimensional images.

There are as many 3D printing processes as there are 3D printer manufacturers (almost by definition, since each one builds a business around ownership of patents for their process). One type of 3D printer shoots a laser into a bath of a polymer that hardens where the laser hits the surface, building up a part in layers by using the laser to plot successive cross-sectional slices as the emerging part lowers into the bath. Another works like a 3D glue gun, squirting molten plastic from a movable nozzle into a chamber that's slightly cooler than the melting temperature, so that it sticks wherever it's deposited.

Yet another process uses an ink-jet printer head to squirt droplets of a liquid binder onto a fine powder, selectively fusing the powder where the printed droplets land. A part is built up by repetitively spreading and fusing layers of the powder. Just as two-dimensional pictures are printed with cyan, magenta, yellow, and black inks, colored three-dimensional parts can be produced by using corresponding colored binder liquids. Another advantage of this process is that the unfused powder serves as a support for partially completed components before they're connected to their structural supports, such as the overhang of an awning. Here's a hello-world that truly says hello to the world, printed this way:

This is what it looked like halfway through the printing process:

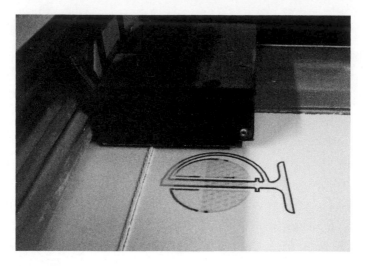

The unfused powder bed surrounds the partially printed part, and if you look carefully you can see where the printer has passed over the new powder layer. The cross section of what will become the shaft passing through the axis of the globe is apparent in the image. The frame will emerge as a single continuous piece, supporting the revolving globe that's being built around it. This kind of connected, nested structure would be impossible to make by conventional machining.

For all the obvious appeal of being able to print almost anything, 3D printers currently have a number of significant limitations. They're priced more like mainframes than printers, and what they do should more accurately be called not-so-rapid prototyping. Build times for a complex part can range from hours to days.

At the trade shows where machine tools are displayed and sold, rapid-prototyping machines occupy an odd kind of technical ghetto. The problem is apparent in the words "rapid prototyping." If one's business is making lots of things, then making a prototype is no more than a means to mass production rather than an end in itself. Three-dimensional printers are seen as filling a niche in product development, but the focus is on the really rapid machines that actually make the products and hence the money. If, however, the prototype *is* the product and the goal is to make and use just one, then you most likely can't afford to buy a 3D printer.

Another limitation of 3D printing is that the beauty of the parts they produce is truly skin-deep—the insides don't do anything. After building a part it's still necessary to add any active components that are needed for an application. If the globe is to turn automatically, a motor must connect to it. To light up cities, LEDs and wiring have to be installed. To show the time of day on the face of the globe, a circuit board would need to be added.

The final frontier in rapid prototyping is to introduce functional as well as structural materials, in order to print complete working systems. Powders and plastics that conduct electricity can be used to print wires, there are printable semiconductors that can be used to deposit logic circuits, motors can be made with magnetic materials, combinations of chemicals can store energy, and so forth. Printable inks containing each of these types of materials have been developed and demonstrated in the laboratory. Integrating all of them into a printer is the most promising route toward making one machine that can make anything. The joke about a student graduating when his or her thesis can walk out of the printer follows from that goal—the student must be able to print the text, the structural elements, the actuators for motion, the control systems for logic, and a supply of energy.

Beyond printing, the ultimate inspiration for additive fabrication is biology. Babies, after all, are grown rather than carved. The molecular machinery in a cell is based around the additive assembly of proteins by other proteins. I'll return to the significant consequences of this simple observation in "The Future."

Building Models

Designers design things, engineers engineer them, and builders build them. There's been a clear progression in their workflow, from high-level description to low-level detail to physical construction. The work at each stage is embodied in models, first of how something will look, then of how it will work, then of how to make it. Those models were originally tangible artifacts, then more recently became computer renderings. Now, thanks to the convergence of computation and fabrication, it's possible to convert back and forth between bits and atoms, between physical and digital representations of an object, by using three-dimensional input and output devices that can scan and print objects instead of just their images. These tools are blurring the boundary between a model of a thing and the thing itself, and are merging the functions of design, engineering, and construction into a new notion of architecture.

Many different domains have had "architects," people responsible for the overall shape of a project. All these architects are starting to use the same kinds of rapid-prototyping tools, both hardware and software. Better ways to build a model and model what's built offer this new kind of architect—whether of a house, a car, or a computer—an opportunity to work not just faster and cheaper but also better, tackling tasks with a level of complexity that would be beyond the barriers imposed by the traditional division of labor between developing and producing a design. The introduction of automation into the making of models has implications across the economic spectrum.

Frank

Frank Gehry is the architectural equivalent of a rock star. The building he designed for the Guggenheim Museum in Bilbao, Spain, was a rare crossover hit, simultaneously transforming the popular perceptions of the nature of a building, a museum, a region, and even of the whole practice of architecture. An enormous number of people who wouldn't have thought much about buildings have been transfixed by the expressive form of this commanding structure (which also happens to house some great art). Few of Frank's fans realize, however, that the way his buildings are designed and constructed is even more revolutionary than how they look. They can appear to be as random as a crumpled-up piece of paper, and that's in fact exactly how they start out.

Gehry began his architectural practice in Los Angeles in 1962. Starting with single-family homes, he became interested in incorporating more expressive materials and forms in architecture. The way he worked was fundamentally physical, based on manipulating models made out of easily modified materials—cardboard, plastic, foam, clay, and the like. The models weren't just representations of a design; making the models became the process by which the design itself was developed. How the models looked was inextricably linked to how they were made: Rather than rigid rectilinear forms, the flexible materials most naturally created the fluid curves that conveyed the sense of movement he sought to construct.

Frank Gehry's flowing forms led to a landmark commission in 1989 to work on a giant scale: a sculpture for Barcelona's Olympic Village.

Modeling MIT's Stata Center

He designed a stylized representation of a fish, 180 feet long and 115 feet tall, that manages to be simultaneously organic and enormous. It also doesn't have any straight lines. Every piece in the apparently simple fish was different, turning it into a dauntingly complex construction project. For help with how to turn his tabletop model into this huge reality, Frank Gehry's office urgently contacted his UCLA colleague Bill Mitchell, a CAD (computer-aided design) guru who later became MIT's dean of architecture.

As Frank's architectural practice grew, he resolutely avoided the use of computers. He saw no need for what was not-so-politely referred to as "Nintendo crap." With all due respect to Nintendo, at the time, computers were seen by architects as more like a video game than a serious expressive medium. But a flowing three-dimensional form like the fish could not be adequately captured by the two-dimensional drawings that architects transmit to engineers, and engineers to contractors.

Bill Mitchell responded to the query from Frank's staff by introducing them to the computer modeling tools used in the aerospace and automotive industries rather than existing architectural software. Because the development and tooling expense for producing a new plane or car can reach billions of dollars, it's of some interest to get it right before committing to a design. This has led to the development of sophisticated engineering software environments that can carry a design from conception, through simulated testing, and all the way to production, without ever appearing on paper. Unlike early architectural programs that mimicked a drafting table, engineering software was increasingly able to model all aspects of the physical world—a world that buildings as well as airplanes inhabit.

In 1990 Jim Glymph, with a background in guiding large architectural projects, joined Frank Gehry to bring CAD into the then computerless office. Jim's job was to carry the flexibility of Frank's physical models into equally expressive computer models that could then control production machinery. The absence of computers in Gehry's operation allowed Jim to skip the two-dimensional drawing stage of architectural software and go directly to three-dimensional design tools. But the software needed to be descriptive, not prescriptive, so that it would not eliminate the essential role of model-making in the design process. Instead, the design defined by a physical model was converted to digital data. Three-dimensional scanners recorded the coordinates of the tabletop models such as the one made for MIT's Stata Center. Engineering software then filled in the supporting mechanical structure and infrastructural services, and the resulting files were electronically transmitted to the companies that fabricated the components. Steel frames and skins were cut and bent to shape under computer control, and assembled on the job site like a giant jigsaw puzzle.

This way of working proved to have a number of advantages. Most important, it made it possible to build buildings unlike any that had come before, because they were limited only by the properties of the materials rather than by the difficulty of describing the designs. But it

Digitizing MIT's Stata Center

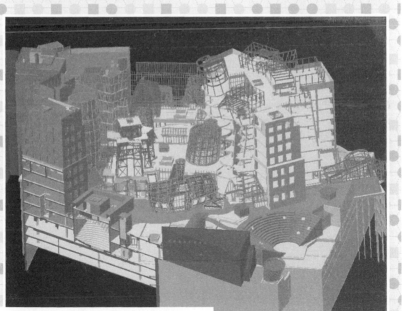

Engineering MIT's Stata Center

Fabricating MIT's Stata Center

Assembling MIT's Stata Center

Constructing MIT's Stata Center

also turned out to be faster and cheaper to build this way. The Barcelona fish went from preliminary design to construction in just six months, beating the planned construction schedule and budget. There was no uncertainty for the contractors because they didn't have to interpret anything; they received a specification with the precision required by a machine rather than the ambiguity accepted by a person. All of the padding that architects, engineers, and contractors built into their schedules and budgets to cover the unpredictability of their communications was removed, because they were all sharing the same computer files rather than regenerating drawings at each stage.

Ultimately, the introduction of engineering design and production tools into architecture challenges the whole division of labor between designing and constructing a building. Once Frank has made a model, he effectively pushes a one-hundred-million-dollar print button to transmit the design all the way through to construction. The real creative work is done by his hands; the rest is relatively automated. If there was a 3D printer as big as a building, it could truly be automated to produce a desired structure, and there are now serious efforts aimed at doing just that.

Larry

Bill Mitchell's collaboration with Frank Gehry's operation was carried on by one of Bill's former students, Larry Sass, who went on to join MIT's architecture faculty. Larry comes from a new generation of architects who grew up comfortable with computers. He mastered programming as a professional skill, and sees computers as an expressive medium rather than just a tool in a workflow.

Larry was interested in the inverse of the way Frank worked. Rather than turning physical models into digital data and then back into components to be pieced together, Larry wondered whether rapid-prototyping processes could convert a computer model of a building into a table-top prototype in a way that reflected its eventual pconstruction. Instead of elaborately emulating the appearance of everyday materials on an enormous scale, he sought to develop designs aimed at automated assembly. In so doing, Larry saw an opportunity for architecture to be responsive to the needs of the poorest rather than the wealthiest members of society.

Larry looked at the design of simple houses. Much like the way my children, Grace and Eli, made their play structures on a lasercutter (described in "Birds and Bikes"), Larry designed models of houses using 2D press-fit panels that could be laser cut out of cardboard and snapped together to produce a 3D prototype. Unlike a toy, though, these models include the load-bearing elements needed to support a full-size structure.

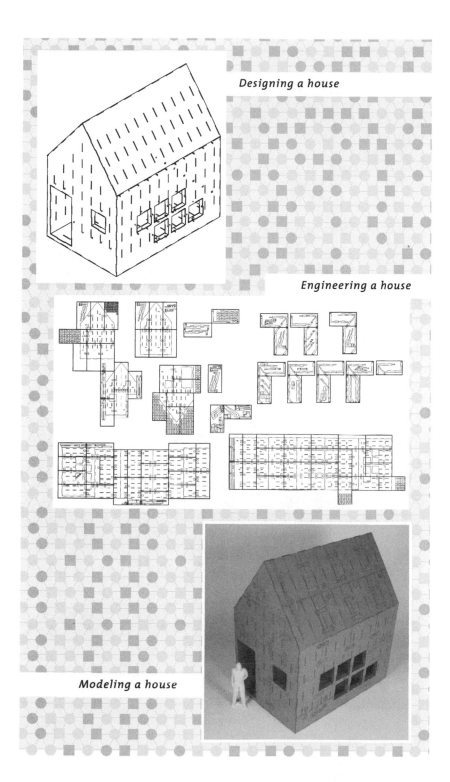

Designing a house

Engineering a house

Modeling a house

Fabricating a house

Constructing a house

Once the tabletop model looked and fit right, the very same design could be scaled up and produced at full size using a computer-controlled routing machine and readily available 4 x 8 foot plywood panels. A router is somewhere between a milling machine and a laser cutter, moving a rotating cutting tool over an enormous bed to plot out the life-size parts. These fit together in exactly the same way they did in the tabletop model, though now the model makes a real house, ready for final sealing and painting.

At my local home products store, a 4 x 8-foot plywood sheet costs about twenty dollars. One of Larry's designs might use one hundred such panels, corresponding to just two thousand dollars in materials for a house. All of that could fit compactly into one shipping container, ready for assembly where it's needed. And because of the computer-controlled cutting, each house kit could vary to respond to the needs of its occupants and the site.

Around the world there are currently a number of other processes under development for house-scale "printing," pumping concrete like a 3D printer, and snapping together supports like a kid's construction kit. Beyond constructing houses faster and cheaper, these projects promise to produce structures that are more responsive to the needs of their occupants, because the most inflexible construction material of all is a stack of static construction documents. By carrying the same digital description of a model from a computer screen to a tabletop prototype to a full-size structure, machines at each step can talk to one another about the dimensions so that people can talk to one another about what really matters: the design.

Etienne

Frank Gehry pioneered the use of CAD tools to turn tabletop prototypes made out of everyday materials into buildings. Larry Sass is using CAD tools to make buildings out of everyday materials. Etienne Delacroix is using CAD tools to make computers out of everyday materials.

Etienne was trained as a physicist, then came to the conclusion that his experience doing science could form a foundation for doing art. He spent fifteen years as a painter, working out of a studio in Paris. Then, in 1998 he came as a visiting scholar to MIT, where he began experimenting with bringing the approach of an artist to the use of technological tools. He made software sketches of painting programs with user interfaces aimed at artists rather than computer scientists. And he started thinking about how to extend the expressiveness of an artist's studio into the domain of computer hardware.

That thought led him on a nomadic journey, ending up in 2000 in Uruguay and Brazil, where he found a fertile intersection of technological, cultural opportunity, and need. In a series of workshops, he started teaching engineers how to work like artists, rather than the more common goal of teaching artists to use the tools of engineers. Like Frank Gehry, Etienne sought to retain the hands-on approach of an artist in a studio, directly manipulating materials. But unlike Gehry, Etienne's materials were electronic.

Etienne started with the mountains of technological junk that are piling up around the world, in poor as well as rich countries. He chopped discarded computers and consumer electronics to bits. Chips and components were desoldered from circuit boards and sorted. Like a good scavenger, he let nothing go to waste—the circuit boards themselves were cut up for use as a construction material, and even the solder was collected for use in making new circuits.

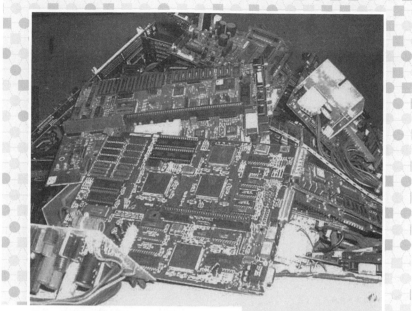

Raw materials for a computer

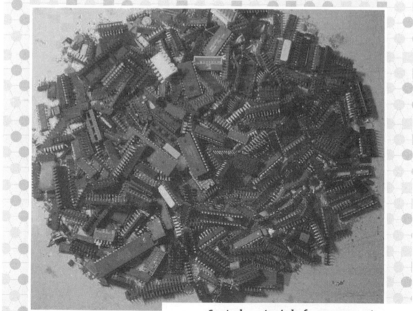

Sorted materials for a computer

Fabricating a computer

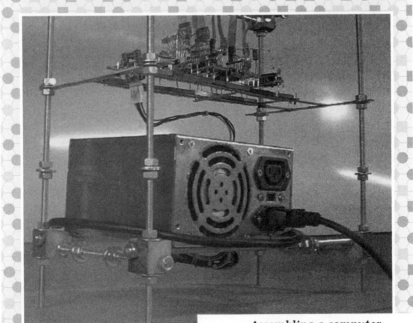

Assembling a computer

The result was a kind of high-tech raw material. Etienne and his students approached it almost like archaeologists, extracting the condensed engineering knowledge that it contained. Having taught his students how to deconstruct modern technology, Etienne showed them how to reconstruct it into new forms. They started with the basics, like power supplies, switches, and lights. As they mastered working with this medium, they progressed through digital logic and microprocessor programming, eventually building complete working computers out of discarded materials.

Hundreds of students showed up for his workshops, somewhere between eager and desperate to gain control over the technological trash around them. In the middle of communities torn by economic, social, and political unrest, Etienne encountered a response exactly like the one I saw at MIT on "How To Make (almost) Anything."

Etienne found that the most difficult technical lesson to teach was imagination. He could see the possibilities lurking within technological junk, but he had a hard time conveying to students how to put the pieces back together short of actually doing it himself. This problem inspired Etienne to turn to the same kind of three-dimensional CAD software that Frank Gehry and Larry Sass were using. He taught his students how to make a virtual version of their studio, freeing them to assemble simulated parts. When they found a good way to put those together, they could then build with the real components, making best use of their available resources to turn trash into treasures.

Etienne's use of CAD tools to model the construction of a computer is literally pushing the boundaries of engineering software. In the aerospace and auto industries, where three-dimensional design tools were developed, the software models the construction of a car or plane. These CAD models contain subsystems, such as the dimensions and connections of a car radio or navigation computer, but don't descend down to the details of individual circuit components. The contents of the subsystems reside with the vendors that produce them. But to simulate remanufacturing discarded digital hardware, Etienne needed to

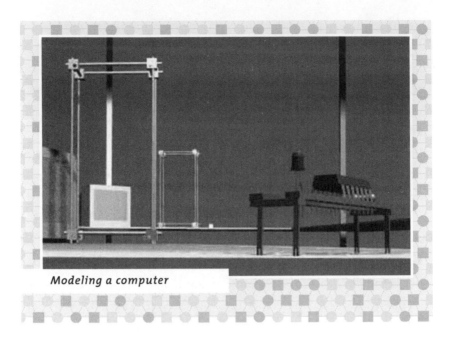

Modeling a computer

open the boundaries between subsystems in order to simultaneously model the electrical and mechanical components, their circuit and structural connections, and the physical and software architecture that holds it all together. In current engineering practice those design functions might be decomposed over ten programs: one for 2D design, another for 3D design, a program to draw circuits, another to lay out printed circuit boards, one to program microcontrollers, a different one to program microprocessors, a program for generating manufacturing toolpaths, one for modeling the mechanical forces, another for modeling electromagnetic radiation, and one more for airflow and heat transfer. A current frontier in the development of engineering software is the integration of all those levels of description into a single environment that can span all the different elements of a design such as Etienne's.

Frank, Larry, and Etienne are all pioneers in exploring the opportunities afforded by turning physical objects into digital data, and vice versa. A mathematical specification, a graphical rendering, a tabletop prototype, and a full-size structure can now contain exactly the same information, just represented in different forms. The intersection of 3D scanning, modeling, and printing blurs the boundaries between artist and engineer, architect and builder, designer and developer, bringing together not just what they do but how they think. When these functions are accessible to individuals, they can reflect the interests of individuals, whether architecting an art museum made to look like a tabletop model, low-income housing made like a tabletop model, or a computer made by modeling recycled refuse.

Description

Personal computing hardware requires software programs in order to do anything. Likewise, personal fabrication hardware requires a description of a design in order to be able to make anything. Those designs could be bought online and downloaded like the songs that go into a portable music player, but the promise of personal fabrication goes beyond consumption to invention. This entails the use of software for computer-aided design (CAD). The purpose of these design tool programs is to help capture what might be an elusive conception of something and convert it into a precise specification for construction.

Physical fabrication tools don't (yet) have an undo button; once an operation is performed on an object it's hard to retract it. At best, CAD software assists with the development of an idea by providing an environment for easy experimentation before fabrication. At worst, CAD software can be infuriatingly difficult to use and impose constraining assumptions about what will be made, and how.

The most affordable, and in many ways versatile, engineering design tool costs around ten cents: a pencil. This often-underrated instrument has a number of desirable features. Its user interface is easy to understand, the response to commands is immediate, it's portable, and it creates high-resolution images. None of the more advanced software tools to come in this chapter can match all of that. For this reason, doodling is an invaluable ingredient in advanced design processes. It can help develop concepts before they get frozen by the vastly more capable but less flexible computerized tools. And sketches can be scanned into a computer as a starting point in the design process, or be sent directly to a fabrication machine as a 2D toolpath.

Beyond ten cents, it's possible to spend thousands and even millions of dollars on design software and the computers to run it. This rarely buys more inspiration. Quite the contrary, actually—the learning curve of high-end engineering design tools can feel more like walking into a cliff rather than up a hill, with an enormous amount that needs to be learned before it's possible to do anything. What serious CAD tools do bring is the ability to manage much more complex projects. As the number of parts (and people) involved in a design grows, keeping track of what goes where, and who is doing what, becomes an increasingly onerous task. Industrial-strength software goes beyond design to manage workflow, from the initial concept all the way through to generating lists of parts to order and instructions to control the production machines, along with controlling the marketers and lawyers and everyone and everything else involved in the life cycle of a product.

To see how this process works (the design, not the lawyer part), let's start with the ten-cent solution:

This hello-world sketch took just ten seconds to produce, and in making it, I had available a choice of fonts and line styles limited only by my nearly nonexistent drawing ability.

The problem with drawing on a piece of paper comes when it's time to make changes, or to express more complex relationships in a drawing. There are just two operating modes: adding marks with the pencil, and removing them with the eraser. Drawing software adds the ability to constrain and modify as well as add and remove marks.

Bitmap- or pixel-based drawing programs manipulate an image just as a piece of paper does, by recording how much electronic ink is stored in each part of the page. *Bitmap* refers to the representation of an image as an array of data bits; a pixel is a *pic*ture *el*ement in such an image, where turning the *c* into an *x* saves the word from sounding like a deli food. Pixel programs are used for expressive things like illustrations, but are less useful for technical drawing. This is because a pixel-based hello-world

Hello World

is made up of many dots:

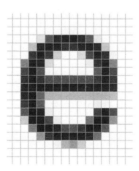

Once these dots are drawn, they no longer know that they're part of an "e"; the only way to change the letter is to erase it and start over.

In contrast to drawing with pixels, a vector-based drawing program effectively saves what *you* do rather than what the piece of paper does. These programs store a line by recording the location of the end points, thereby defining a vector, and more generally they store shapes by recording the points used to define the shapes, such as the center and radius of a circle. This allows an object to be later moved or modified by changing its reference points. And since the shapes are defined by mathematical functions rather than dots, they can be drawn with as much detail as is needed for the output from the CAD environment, whether it is being used to cut steel or place atoms. Most important of all, these shapes can be combined with others to build up more complex structures.

A vector hello-world looks the same as a pixel hello-world from a distance:

Hello World

but it's now possible to zoom in

without losing resolution because the letters are made up of curves rather than dots. Here are the points that define those curves:

Since mathematical functions define the shapes of the letters, mathematical mappings can be applied to them. One particularly useful set of editing operations is the Boolean transformations, named after the mathematician George Boole, who in 1854 formulated a mathematical representation of logic. For example, we could overlay a base rectangle on our hello-world:

and then use Boolean operations to add (top image, below), subtract (middle image), or intersect (bottom image) them:

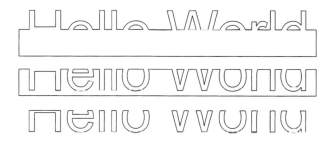

After pixels and vectors, the next distinction in computer drawing programs is between those that can work with 2D shapes and those that work with 3D solids. Just as there are 2D fabrication tools, like a laser cutter, and 3D ones, like an NC mill, there are CAD programs targeted at 2D and at 3D design. In a 3D CAD program, we can draw our hello-world again,

Hello World

but now rotate it,

and then extrude it to form 3D letters:

Once again, we can introduce a pedestal, now three-dimensional,

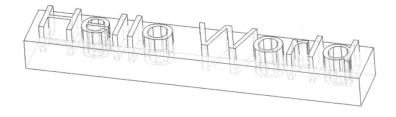

and then use 3D Boolean operations to add (top image, below), subtract (middle image), or intersect (bottom image) the letters with it:

To make the last image easier to interpret, the computer simulated the light and shadow by using a mathematical model of a lamp illuminating the solids. This process, called *rendering*, is an important part of developing design.

Beyond the transition from two to three dimensions, the real aid in computer-aided design comes in programs with a set of capabilities loosely called *parametric*. These programs specify the essential aspects of a design in a way that lets them later be modified. Let's say you're designing a spaceship. It would take an awful lot of mouse clicks to draw in every nut and bolt. After you're done, if you find that the bolts are not strong enough to hold the spaceship together, you would then need to laboriously change them. If instead you define an archetypal bolt, with its diameter specified as an adjustable parameter, and then if this master bolt is duplicated all over the design, then changing that single parameter will modify all of the instances of the bolt. Furthermore, you can define a bolt hole to have what's called an *associative relationship* with a bolt, in this case leaving a desired amount of space around it, and the nut and bolt can be grouped together to form a hierarchical assembly, so that now changing the original bolt automatically enlarges the hole around it, and those changes carry through to all of the nut-and-bolt assemblies duplicated in the design.

As an example of parametric modeling, here's a plate with some holes:

These holes were made in a parametric CAD program by drawing a cylinder, making three copies of it, moving the cylinders to the corners where the holes are to go, and then doing a Boolean subtraction of all of them from the plate to create the holes. Because the program remembers these operations and the relationships they defined, changing the diameter of the original cylinder automatically updates all the holes:

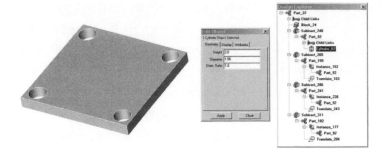

Now, it would be rather awkward to discover that a spaceship's bolts were not strong enough once you're on your way to Mars, so still more capable CAD programs add the ability to model the behavior as well as the shape of a part. Using the mathematical technique of finite element analysis (FEM), it's possible to check how a design will respond to most anything: forces, vibration, heat, airflow, fluid flow, and electric and magnetic fields. Finite element analysis works by covering a structure with a mathematical mesh of points

and then calculating how those points move in response to applied forces, here pushing up in the middle:

The shading corresponds to the calculated local strain (displacement) in the material, with lighter areas moving more. This calculation provides qualitative insight into where the part should be strengthened, and quantitative information as to whether the strain is so large that the material will crack. Here it can be seen that the area around the top of the *W* is a likely failure location.

Even with all the enhancements, these examples of CAD tools still share a particularly serious limitation: they fail to take full advantage of the evolution of our species over the past few million years. The human race has put a great deal of effort into evolving two hands that work in three dimensions, with eyes to match, but the interface to these design tools assumes only one hand that works in two dimensions (dragging a mouse around a desktop). A frontier in CAD systems is the introduction of user interfaces that capture the physical capabilities of their users. These range from knobs and sliders that can control a three-dimensional graphical view, to data gloves that can convert the location of your hands into the location of on-screen hands that can directly manipulate a computer model, to whole rooms with displays projected on the walls that make it possible to go inside a computer model. The most interesting approach of all is to abandon the use of a computer as a design tool and revert to the sophisticated modeling materials used in a well-equipped nursery school, like clay:

An image of the clay can be sent to a two-dimensional cutting tool, here using the contours around the letters as a toolpath for a sign cutter to plot them in a copper sheet:

The same file could be cut in acrylic on a laser cutter, or in glass or steel on a waterjet cutter. Alternatively, a three-dimensional "picture" of the clay could be taken by using a 3D scanner:

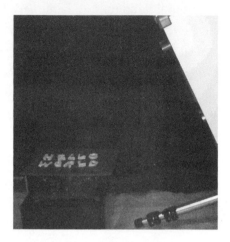

There's a laser in the scanner on the right that's moving a spot back and forth over the surface of the clay. The scanner uses a camera to watch where the laser light is reflected back to it, and from those measurements it can deduce the shape of the surface that's scattering the laser beam. The result is a computer model of the clay:

These shapes would have been difficult to define from scratch as a mathematical model, but once scanned and imported into a CAD environment they can be manipulated like any other object: scaled larger or smaller, bent around a surface, or merged with other components by Boolean operations. The resulting model can then be output on a fabrication tool such as a 3D printer or an NC mill, effectively converting the clay into plastic or steel, with enhancements added through the CAD editing. Used this way, the computer assists with the development of a design but it isn't expected to be the medium in which the original inspiration happens.

In the end, there is no single best kind of CAD. Software design tools are really much like any other kind of tool. No one tool, hardware or software, can do all jobs equally well. Just as a serious artist works surrounded by a variety of paints and brushes, serious computer-aided design entails working with a collection of programs. Like the tools on a workbench, there are familiar favorites with well-worn ways of working, less-used special-purpose programs, and the occasional custom program needed for a particular task.

Ultimately, the development of universal programmable personal digital fabricators may drive the development of correspondingly universal personal digital design software. Until then, CAD will remain a fundamentally enabling, occasionally exasperating, and altogether essential part of personal fabrication.

Playing at Work

Seymour

The expected division of labor between kids and adults is for the former to play and the latter to work. This is particularly true of their relationship with technology, which creates toys for kids and tools for adults. But the inventiveness of children has led to a historical blurring of the distinction between toys and tools for invention, culminating in the integration of play and work in the technology for personal fabrication. The original inspiration and instigator for bringing these worlds together was Seymour Papert, a mathematician turned computer pioneer.

Seymour first encountered computers by a fortuitous accident. He had met MIT's Marvin Minsky, considered to be the father of the study of artificial intelligence (AI), at a conference in the late 1950s. Their meeting was memorable because they had, without having heard of each other, submitted essentially the same paper to the conference. They were both thinking about how to mathematically model the way the brain reasons, in order to better understand it. Both approached this question by abstracting the essential features of networks of neurons;

the similarity of their approaches led to their joint authorship of a book titled *Perceptrons*, which laid the foundation for generations of study of neural networks. But their meeting also led Seymour in a very unexpected direction when Marvin invited him to come to MIT.

Seymour had been in Geneva, working with the famous Swiss psychologist Jean Piaget. Piaget articulated the "constructivist" approach to learning, recognizing that children learn through experimentation much more than by recitation and repetition. They're really little scientists doing endless experiments, an admirable trait that gets trained out of them in a conventional education based around a rigid curriculum.

In 1963 Seymour traveled to MIT expecting to extend his study of learning by developing better models of artificial intelligence. Seymour arrived, but Marvin didn't: there was no sign of Marvin at the appointed time and place (it later turned out that Marvin had remembered the wrong day for their first meeting). This fortuitous mix-up left Seymour sitting alone in a room with a curious-looking machine that turned out to be one of the first minicomputers, Digital Equipment Corporation's PDP-1, serial number 2. This was the commercial descendant of the earlier experimental TX–2 computer developed at MIT, and as part of the licensing deal the DEC agreed to provide a preproduction model of the PDP-1 for use at MIT.

At that point in the history of computation, even an hour of mainframe time was a precious commodity. But the PDP-1 was sitting idle because there was not yet a demand for individual computer use. No one was sure what it would be good for. Mainframes might be valuable for running accounting systems or tracking inventory, but who wants to do that in their free time?

Since the computer was available, and Marvin wasn't, Seymour started fiddling with it. He learned how to program it, planning to use it to generate optical illusions with random dot patterns as a way to study how human perception works. His success doing that opened his eyes to an even greater possibility: in the programmability of the computer, he had found the ultimate sandbox for kids to play in. If children could get the same access Seymour had to a machine like the PDP-1, they could play with abstract concepts with the same ease that they play with a lump of clay.

There were two problems with realizing this vision. The first was the PDP's machine language that Seymour had to learn to program it. We'll meet machine language in the next chapter; it's about as unfriendly as a language can be. Seymour needed to develop a language that both kids and computers could understand. At that time, the new language of choice for computer scientists at MIT was LISP. A bit of the LISP code for Emacs, the editor I'm using to write this book, looks like this:

```
(if (= (car n) Ø)
   (cons Ø s)
   (cons 1 (cond
        ((not (listp (cdr n)))
        (list 'vconcat (cdr n)))
        ((eq (nth 1 n) 'list)
        (cons 'vector (nthcdr 2 n)))
        ((eq (nth 1 n) 'append)
        (cons 'vconcat (nthcdr 2 n)))
        (t
        (list 'apply '(function vector) (cdr n)))))))))
```

For anyone, like me, whose brain does not work like a computer scientist's, this is a foreign language. But it is based on an elegant model of programs operating on lists of data, which led Seymour to abstract its essence into the LOGO language aimed at kids. LOGO works much like LISP, but unlike LISP, it's written in words that mortals can read.

The second problem was providing kids with access to computers; the PDP-1 wasn't even on the market yet. Seymour solved this by employing the newly developed capability of computers to do more than one thing at a time. Time-sharing made it possible for a remote

Teaching turtles

user in a classroom to dial into MIT and run a program, with the computer switching between multiple users' programs so fast that it appears to any one of them to respond in real time. At the time, this was harder than it sounds; the development of "high-speed" telephone modems that could send data at three hundred bits per second was a cause for great celebration (today's cable modems run at millions of bits per second).

Seymour tried an experiment, connecting kids in an inner-city Boston classroom to remote computers. This worked well with older children. Instead of Seymour's original plan to develop models of AI, he had the kids do it. Just like AI researchers at MIT using LISP, the children used LOGO to explore the consequences of rules they wrote for how the computer should respond to them. But that was too abstract to engage younger kids, below around fifth grade. Because physical manipulation is essential to how they learn, Seymour wanted to provide physical devices they could use to interact with the computer.

A staple of early AI research was a robotic "turtle," a little vehicle controlled by an electronic nervous system. Seymour developed a turtle that could connect to a remote time-sharing computer. A program's output could be embodied in how the turtle moved and reacted to its surroundings. Along with the turtle a number of other physical interfaces were developed, including a box of control buttons, and a device that could read instructions on cards. (The latter was developed by a precocious undergrad, Danny Hillis, who went on to become a precocious adult architect of supercomputers.)

These capabilities were far ahead of what was available to the typical corporate computer user, who had to wait years before this kind of real-time computer interaction became available in mice, pens, plotters, and other input-output devices. For kids, the innovation in the physical form of their computer interfaces paused when computers developed to the point that turtles moved onto the screen as computer graphics (around 1970). This was not as compelling as an (artificially) alive turtle moving in the real world, but it made LOGO programming available to anyone with access to a computer. Versions of LOGO migrated from minicomputers to the emerging personal computers, from the TI–99 to the Atari 400 to the Apple II to the IBM PC.

While this was happening, a new collaborator appeared, thanks to the BBC. Kjeld Kirk Kristiansen, the head of LEGO, saw a BBC news program about the turtles, and is reported to have said, "Hey, there's someone who thinks like us!" At the time, LEGO was not yet the icon it's become. LEGO was started by his grandfather, Ole Kirk Christiansen, an out-of-work carpenter eking out a living in Denmark in the 1930s making wooden toys. After the interruption of World War II, Ole Kirk decided he liked making toys so much that he went back to it. One of his products was a wooden truck. When moldable plastics came along, he used them to add bricks as a load for the truck. Sales of the

Controlling computers

Programmable brick

truck took off, but it was bought for the bricks, which were more fun than the truck. The rest is business history.

If a computer could connect to a turtle then it could connect to LEGO bricks, allowing kids to add functional behaviors to the physical shapes they were making. Seymour's successors at MIT, led by Mitch Resnick, worked with LEGO on developing kits to interface LEGO sensors and motors to programmed LOGO worlds in PCs. This was promising, but playing with the bricks still required a desktop computer.

Meanwhile, computing continued its evolution from microprocessors powering PCs to even smaller microcontroller chips that were complete simple computers that could be embedded in products. Mitch and his colleagues Fred Martin, Brian Silverman, and Randy Sargent used a microcontroller to develop a LEGO/LOGO controller about the size of a child's wooden block. This became the model for the Mindstorms robotic construction kit (named after a book by Seymour). That group, along with Bakhtiar Mikhak and Robbie Berg, continued this evolution, shrinking the processing down to the size of a single LEGO brick.

Kids loved this, in a number of often unexpected ways. In an early test in a fourth-grade class in inner-city Boston, the boys quickly caught on to the computing bricks, using them to make all sorts of vehicles. The girls weren't so keen on the computing part, choosing to stick with the original bricks to design houses. But they were eyeing what the boys were doing, and discreetly started adding features to the houses, like lights that could be turned on and off. Then they started programming patterns for the lights, and from there started inventing robotic appliances. They were hooked, and didn't even realize that they were learning how to design engineering control systems.

David

In 1999 I wrote a book, *When Things Start to Think*, on the emerging technology and implications of computing moving out of conventional computers and into everyday objects, from LEGO bricks to footwear to furniture. I was pleasantly overwhelmed by the detailed letters I received from grown-up engineers, discussing their own inventions, past, present, and future, ranging from eminently practical to utterly impossible. One of these stood out as being particularly well written. It began with a thoughtful critique of the book, then veered off in an unexpected direction:

> True or False:
> A school bus in 2045 will:
> a)   have sensor-controlled gyroscope stabilizers.
> b)   be able to detect rule-breakers and their names, then report them to the driver.
> c)   have thought-controlled windows.
> d)   store in memory frequented destinations (such as the indigenous school), then drive the bus automatically (Now there's an AUTOmobile!) to those destinations.

Intrigued, and puzzled, I read on to find closing observations including "The future of the future looks bright." I then nearly fell out of my chair when I read the signature: "Cordially, David (Just turned 8 years old, about to enter 9th grade)." Not sure if this was serious or an odd kind of joke, I wrote back a tentative reply, which was followed apparently instantly by a seven-page letter from his mother, Athena, which roughly translated to "Help!!!!" She was home-schooling David because regular schools couldn't keep up with him. He was indeed only eight years old, and her problem was that he was ready for college and she wasn't sure what to do with him.

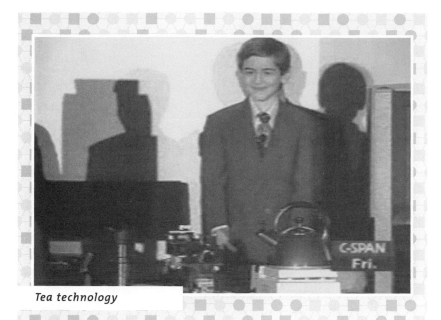

Tea technology

This began a lively correspondence with both David and Athena. Through our exchanges I learned that David had been saving up to buy a Mindstorms kit, because his ideas for inventions were racing far ahead of his ability to realize them. I arranged for him to get one, which he quickly devoured, mastered, and put to use in projects.

I thought of David again when that same year the White House asked me to put together a panel on the future of science and technology for the millennium celebrations. If our future is our children, and our technological future lies in their inventions, who better to give a peek at that than an eight-year-old? Along with an assortment of older students and grown-up scientists I added David to the panel, unwittingly making a terrible mistake.

All sorts of impressive technology was on display, from real-time atomic manipulation to embedded intelligence in paper, clothes, and furniture. David brought a Mindstorms invention he made for his mother to address her frustration with her cup of tea being either too hot or too cold. To ensure that her tea was just right, he made a machine that mixed in cold water, stirred the tea, and announced when it had cooled down to the ideal temperature.

The mistake I made was not putting David at the end of the program: He stole the show. After him, it was hard for anyone to present anything else. David's presentation was smart, funny, relevant, and moving. He used his tea server as a springboard to discuss how broader access to the tools for technological development could help humanize technology, better meeting society's needs by involving individuals in posing and solving their most pressing problems. The future of the future is indeed bright in the hands of young inventors like David.

Ken

The biggest surprise when LEGO started selling Mindstorms was who bought them. Of the adults leaving toy stores with kits, about half turned out to be getting it for themselves rather than for children. Engineers who might work as small cogs in big projects were attracted by the opportunity to build complete systems themselves. And, even more surprising, they weren't doing this just for fun. I first came across this when I was visiting a colleague doing physics research at a lab that had recently won a Nobel prize; he was happily assembling LEGO components to put together a control system for his latest experiment.

Back at MIT the same thing was happening, as grown-ups adopted as serious tools technological toys developed for kids. One of them was Ken Paul, a process engineer from the United States Postal Service. He was based on campus to investigate applications of emerging research results for USPS operations, which were expected to include things like putting radio tags in envelopes to help better route the mail.

As Ken Paul went about his business he did something a bit like what the girls did when the first computing bricks came into the classroom, when they discreetly watched what the boys were doing and

Serious work

then started adding intelligence to the playhouses they were building. Kenny eyed the kids playing with the new generations of computing bricks that Mitch, Bakhtiar, and their colleagues were developing, and he thought that what they were doing looked like more fun than what he was doing. More seriously, he wondered if they could be used to help model billion-dollar infrastructural decisions.

The USPS operates on such a large scale that in its acquisition and operations budget, there's a dangerous gap between funding an exploratory paper study and committing to a large-scale deployment. It's not possible to build a football-field-sized mail-sorting facility as a prototype to evaluate and optimize the design, so there's a leap of faith to go from a design to actually building the facility. Unless, of course, the football field can fit on a tabletop. Ken started working with Bakhtiar Mikhak and Tim Gorton at MIT to make LEGO-scale models of these big installations, which looked like toys but operated according to the rules of their real counterparts.

He was hooked. As he progressed, with some trepidation we began planning a visit to Washington, DC, to present the project to USPS management. Ken worried that the project might appear to be pretty frivolous when he was seen playing with toys.

We trooped down to Washington, arrived at the executive conference room, and set up his system. The management team then filed in, looking suitably serious. Eyebrows raised when they saw what looked more like a playroom. But as Kenny explained what he was doing, and then did a demonstration, their eyes widened. Managers who spent their days on the receiving end of thick reports asked whether even they could use a tabletop prototyping system to try out what-if scenarios themselves instead of paying high-priced consultants to do the analyses.

The importance of Ken's demonstration went far beyond the particular mail-handling application he had chosen. Just as spreadsheets became a killer application for PCs because they allowed managers to make models of their financial flows, the computing bricks allowed the USPS managers to model physical flows of materials and information. Business executives as well as kids like hands-on interfaces, immediate feedback on their actions, and the ability to work together in groups to solve problems. There's very little difference in the technology for serious work and serious play.

Amy

Amy Sun was a prized defense engineer, doing critical work for key government programs on earth and in space. Outside of her day job she had a consuming interest in engaging kids in science and technology, ranging from helping with school outreach programs to building battling robots. I met her in 2002 when she joined my colleagues Professor Ike Chuang, CBA's program manager Sherry Lassiter, and MIT undergrad Caroline McEnnis, in setting up our first field fab lab, in India at Kalbag's Vigyan Ashram (a rural science school described in "Making Sense").

Before going to India, Amy stopped by MIT to help get everything ready, and in the best tradition of incoming MIT students, she promptly took over the lab. She quickly mastered the larger on-campus versions of the tabletop rapid-prototyping tools going into the field, using them to make parts needed for the trip, and much more. Her decision after the trip to become a grad student at MIT followed more as a statement than as a question; she (correctly) assumed that MIT's application formalities were just a detail, given the obvious match with her abilities and interests.

After a delay while she handed off her projects, she interrupted her previously scheduled life to come to MIT as a student. Along with her research on campus, which entailed shrinking LEGO bricks down to a millionth of a meter in order to assemble microscopic three-dimensional structures, Amy was soon back in the field helping start fab labs in inner-city Boston in 2003 at Mel King's South End Technology Center, and in 2004 in Ghana at the Takoradi Technical Institute.

She created quite a stir when she arrived in Africa; a typical comment that came back while she was there was, "The lady is amazing she got every stuff in her!" Even better, she was such a strong role model as an engineer that boys using the fab lab asked her questions that had perhaps never been asked before: "Is MIT a girls' school? Will they let boys in?"

Amy started off fab lab users in Africa by having them use the laser cutter to make puzzles that they had to assemble into geometrical shapes. This was intended to be a warm-up lesson for the students, but it turned into an even greater lesson for the teachers: The street kids coming into the lab were able to solve the puzzles much more quickly than the older students or adults there.

Ironically, while this was going on one of the older teachers in the school asked why the high-tech tools in the fab lab were being wasted on young children. It was an eye-opening moment for him when he

Teaching technology

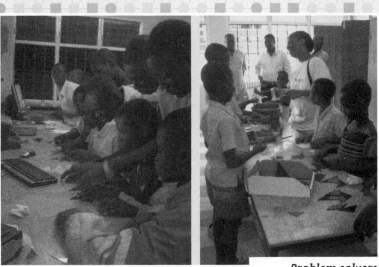

Problem solvers

saw what those kids could do. Not too long afterwards, a local reverend admonished the community to go home and start fasting and praying because "the Bible tells us that we should be telling the children to come unto and instead we have been pushing our children away." He was struck by finding that a resource as valuable as the fab lab could be shared with kids rather than protected from them. That's a lesson that kids of all ages in the lab did understand. A single seat might be occupied by ten bottoms, as young and old, boy and girl, piled on to share a tool, instinctively collaborating and teaching one another.

As work in the fab lab progressed from teaching tutorials to working on real-world applications, it ran into a significant limitation. One of the highest priorities to quickly emerge in the Ghana fab lab was solar energy, seeking to develop machines that could directly harness the abundant power of concentrated sunlight without the cost and inefficiency of converting it to electricity first. This activity required making solar collectors that were not only bigger than the cutting tools in the fab lab but needed to be bigger than the lab itself. But how can a machine make something bigger than itself?

An answer can be found all the way back in Seymour Papert's turtle. The fab lab could make a mobile computer-controlled car that could drive over the material to be used, trailing a pen to plot precise shapes that could then be cut out with simple hand tools. The original turtles provided graphical feedback long before computer graphics were up to the job; this new kind of turtle could do the same for fabrication. A project soon followed to develop a turtle design that could be produced in the fab lab.

This solution to the practical need to plot large structures represents the realization of one of Seymour Papert's original dreams. As exciting as bringing computers and controllers into the classroom once was, he describes the inability of the kids back then to invent as well as use technology as a "thorn in our flesh." Unlike what's possible with the materials in any well-equipped arts-and-crafts area, the turtle provided output from the computer but the physical form of the turtle itself couldn't be changed. Its fixed form put a severe bound on the shape of kids' ideas.

The possibility now of making a turtle in a fab lab is important as an application, allowing the production of structures bigger than the lab. The implications range all the way up to global energy economics in fabricating large-scale solar collectors. But a do-it-yourself turtle is even more important as a process, breaking down the barrier between using and creating technological tools. Like real turtles, one starting design can evolve into many subspecies. The inspiration for such an invention can be play, or work, and, best of all, because it can be done by an individual rather than justified by an organization it's not necessary to try to tell the difference.

Computation

The preceding chapters introduced tools for creating physical forms: designing, cutting, and printing three-dimensional structures. We turn now to the creation of logical functions, the intelligence that breathes life into inanimate shapes. Doing this requires embedding the means for those objects to act and react through programs that define their behavior.

The key component that makes the embedding of intelligence possible is the microcontroller. It is distinct from the microprocessor that is the heart, or more correctly the brain, of your PC. Or even more correctly, it's the prefrontal cortex of the brain, which is where critical thought happens. A microprocessor is just that: a small (compared to a mainframe) information-processing device. All it can do is think; it needs other components to remember, or communicate, or interact. It's a brain that is useless without the body of a PC to provide its inputs and act on its outputs.

Compared to this, a microcontroller is more like an insect. It would be generous to call an ant's rudimentary nervous system a brain; its intelligence is distributed over the length of an ant rather than centralized in one place as an abstract thought process. The nervous system is tightly integrated with the ant's organs, efficiently executing a static set of responses. That fixed program makes an individual ant less adaptable, but what an ant does do it does very well. I doubt that I could get through the first day on the job as a worker ant; because of my superior intellect, I would need too much supervision, take too many breaks, and too quickly lose interest in collecting crumbs.

A microcontroller is much smaller and cheaper than a microprocessor, but it also does more with less. In addition to having a core computer processor, there can be permanent memory for storing programs, converters between analog and digital signals to interface with input and output devices, timers for real-time applications, and controllers for external communications. Because it doesn't do any one of these functions as well as a dedicated chip, all of this can fit into a tiny package costing a dollar.

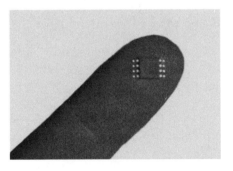

In the fullness of time, a printer using inks made of electronic materials (semiconductors for transistors, conductors for wires, insulators for isolation) will be able to print a microcontroller. In the shortness of time, microcontrollers already come pretty close to being a consumable much like a printer's ink. Because they're so cheap, flexible, and easy to stick into or onto almost anything, they can be used as a raw material for rapid prototyping.

What makes little microcontrollers so powerful is their speed. The fastest are based on what's called a *RISC* design. This doesn't mean that they're dangerous: RISC stands for *r*educed *i*nstruction *s*et *c*omputer. Any computer has a set of basic instructions that it understands. In a conventional computer there can be hundreds of these, requiring many operations (and hence time) to interpret them. In a RISC processor this list is pared down to just the most essential ones, simplifying the hardware and allowing the execution of any of the instructions to take place in a single cycle of the digital clock that orchestrates the operation of the processor.

This means that even a relatively modest RISC microcontroller can execute instructions faster than a million times a second (a microsecond). That's important because it's much faster than many of the things you might like a microcontroller to do. The smallest time difference that a person can perceive is about a hundredth of a second, or ten thousand times slower than a typical microcontroller clock speed. This means that such a processor can appear to react instantaneously while actually being able to do the work of ten thousand instructions. The sound waves of a favorite recording, or the signals used by a telephone modem, vary tens of thousands of times a second, which is still about a hundred times slower than the processor runs. And even the higher-speed signals used in radios or video displays or data networks can be slower than a fast microcontroller. Therefore, all those different functions—audio, video, communications, user interfaces—can be handled by a single general-purpose microcontroller rather than many special-purpose circuits.

To show how that works I'll make a microcontroller say "hello world" (of course), by sending those characters to a desktop computer. The first step is to draw a circuit showing how the required components are connected. Any drawing program could be used, but it is usually done with a program dedicated to drawing schematic diagrams of circuits. Unlike a general-purpose drawing tool, a schematic program keeps track of the available components and how they get connected, so that it's able to retain those connections as parts get moved around in the drawing, and can help spot if incompatible pins are erroneously connected.

Here's a schematic for the hello-world circuit:

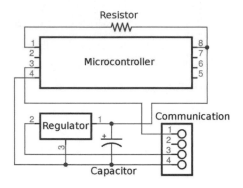

Because the microcontroller uses so little current to operate, about 0.001 amps (a milliamp, written 1 mA), this circuit uses a trick to avoid the need for a battery or power supply: It steals the power. The computer communication interface (to be described in "Communication") uses one wire to send data and another to indicate when data are ready to be sent. The latter is called data terminal ready (DTR). But the act of simply turning on the DTR line provides energy as well as information. A typical DTR line can supply about 10 mA of current. In this hello-world circuit, a device called a regulator turns the DTR line into a steady power supply for the processor. The other components in the circuit are a capacitor that smooths out variations in the supply, a resistor that tells the processor to keep running once it turns on, and a connector for the communications cable.

The next step in making this circuit is to use another program to lay out the physical arrangement of the components. This starts from a so-called "rat's nest" of connections from the schematic program, which are then routed based on the desired layout of the circuit board and the constraints of the sizes and connections of the components:

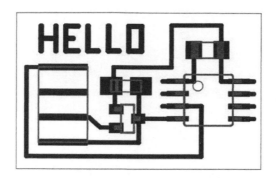

For extra credit, wires were added to spell out the letters *HELLO*.

These traces become what's traditionally called a *printed circuit board* (PCB), although such boards are most often made by etching rather than printing. The most common commercial process uses an optical mask to expose an image of the board onto a film layer applied on top of a copper layer, which itself is on top of an insulating board. The unexposed film is then washed away, and a chemical solution etches away the exposed copper, leaving behind the desired traces. For more complex circuits, this procedure can be repeated ten or more times to build up multilayered boards out of individual circuit layers. There are processes that directly print wiring, eliminating the need to expose and etch the boards, but their use has been limited because expensive materials such as silver inks are required to directly deposit conductors good enough for electronics.

To avoid creating a global revolution in the local production of chemical waste, the fab labs use a subtractive machining process instead of chemical etching. An inexpensive tabletop high-resolution milling machine can mechanically remove copper around the desired traces. Machining is slower than etching a board, which happens simultaneously all over the board rather than sequentially following a toolpath, but the great virtues of this process are that the only waste is the small amount of removed material, and it's just as easy to make every board unique as it is to repeat the same design over and over again.

Alternatively, if the circuit needs to be flexible, or mounted directly onto something else, the traces can be cut out on the vinyl cutter and applied to a substrate:

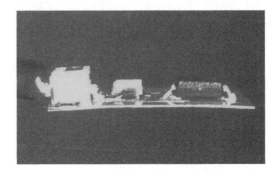

However a board is made, the next step is to put the parts onto it. There are two kinds of components for making electronic circuits: "through-hole" parts that mount via leads going through holes in the board, and "surface-mount" parts that attach directly onto the surface of the board. Surface-mount parts are more recent, smaller, cheaper, and they simplify board fabrication, so they are replacing through-hole parts everywhere but where the added strength of leads in mounting holes is needed. Using a 1/64-inch (0.0156-inch) end mill in a milling machine, it's easy to cut away the copper to produce pads and traces finer than the 0.050-inch spacing that is common for connecting to surface-mount parts:

Attaching the components is called *stuffing* the board. In mass production, "pick-and-place" machines do this. These are high-speed robotic hands that remove parts from reels and stick them onto the board with a solder paste that is then melted to make the connections. Pick-and-place machines are fast, expensive, and require a great deal of setup before they can assemble anything. But there's a trick that makes it possible to manually solder surface-mount parts without requiring a pick-and-place.

Solder is a mixture of metals that melts at a temperature lower than the melting temperature of any one of them alone, typically about 400° Fahrenheit. A soldering iron, a hot-air blower, or a small oven can apply that heat to a component on the board. Once the solder melts, it will "wet" materials at that temperature as a drop of water sticks to a surface. This means that if solder is fed onto the heated leads of a chip resting on circuit board traces, it will preferentially stick to the thermally conducting leads and traces rather than the insulating gaps between them. The solder just needs to be applied in the neighborhood of the desired lead and it will end up in the right place. In this way it's possible to use simple handheld tools to solder chips with features finer than the resolution of the tools or the hands holding them.

Once the board is stuffed, the next step is to write the program to teach the microcontroller to say "hello world." There are many different ways to do that, based on how much work the programmer versus the computer does. At one extreme, the program can be as simple as:

```
while True:
    print 'hello world'
```

This is written in my favorite high-level language, Python. The line "while True" means that what follows should be executed over and over until the computer equivalent of hell freezing over, and the line "print 'hello world'" does just that. When this program is run, it repeatedly prints the text "hello world."

This program is simple to write, but complex for the computer to execute. Python is an interpreted language, meaning that what's really running is another program that converts Python statements into lower-level instructions that the computer can understand. To do this, the computer requires more memory to run both the Python interpreter and the hello-world program to be interpreted, and the effective speed of the computer will be slower because of the time consumed by doing the interpretation. Nevertheless, interpreters have been written to run in even the tiniest mircontrollers, notably for the LOGO and BASIC languages.

The need for an interpreter can be avoided by using a lower-level compiled language, which shifts some of the work from the computer to the programmer. A program called a compiler converts these kinds of programs into computer instructions that can be executed directly without needing any other programs. In theory, a program written in an interpreted language could be compiled, and a compiled language could be interpreted. In practice, that's not usually done because the structure of compiled languages is a bit closer than interpreted languages to how computers rather than people think so that it's easier to write compilers for those languages.

Here's the hello-world written in the compiled language of the original hello-world, C (so-named because it was the successor to a language called B):

```
main() {
  while (1) {
    printf("hello world\n");
  }
}
```

The number 1 replaces the word "True" here because that's how a computer represents the concept, and the "main()" instruction is there because a more complex program might call subprograms from this main one. C compilers exist and are widely used for most every known microcontroller.

The C hello-world program is still remote from what really happens in a microcontroller. Each family of microcontrollers has a native set of instructions that they understand. Each of these "op-codes" (operation codes) corresponds to a primitive operation that a microcontroller can do, such as adding two numbers, turning a pin on or off, or moving a value between memory locations. The job of a compiler is to turn commands written in a high-level language into op-codes. That's a nontrivial task, because one line in a high-level program can correspond to many low-level instructions, and there can be many different ways to map those instructions. Choosing the best way to do that requires an understanding of the purpose of the program.

Rather than attempting to automate the translation from a compiled program to op-codes, it's possible to directly write a program in the instruction set of a particular processor. This results in the smallest, fastest programs. After all, the author of a program understands it best. Here's what the hello-world looks like written in a microcontroller's native instructions:

```
print:
  print_loop:
```

```
            lpm
            mov txbyte,RØ
            cpi txbyte,Ø
            breq return
            rcall putchar
            inc zl
            rjmp print_loop
          return:
            ret
      print_string:
        .db "hello world",13,1Ø,Ø
      reset:
        sbi PORTB, txpin
        sbi DDRB, txpin
        loop:
          ldi zl,low(print_string*2)
          ldi zh,high(print_string*2)
          rcall print
          rcall print_delay
          rjmp loop
```

Each of these cryptic commands represents one instruction. For example, "inc" is short for "increment," which adds 1 to the value stored at a memory location. Here it increments the location named "zl," which points to the next character to be printed. Each line in the C program has been unpacked into all of the primitive operations required to implement it.

Programs written this way are called "assembly language," because they must pass through a program called an assembler before they can be loaded into a microcontroller. For the French philosophers in the audience, the assembly language instructions are a signifier for the real instructions that the microcontroller understands, which are stored as numbers rather than words to save space. The assembler makes this straightforward conversion. These programs that reside in a microcontroller are also called *microcode* because they're the code that actually resides in a micro.

Assembly language programs are easiest for the computer to understand, but hardest for a person to understand. That makes them longer to write, but faster to run. Their power is pointless if one's ambitions do not extend very far beyond printing "hello world." However, if the "hello world" is to appear as a video signal on a TV set, and if a modest microcontroller is used to generate those high-speed signals, then assembly language is likely to be used both for its speed and control over the timing. Since faster processors cost more, use more power, and come in larger packages, trading off that extra programming effort for a smaller, faster program can be a good investment.

After assembling, here's the hello-world program, ready for loading:

```
:020000020000FC
:1000000028C00AE03095089410F4C39A02C0C3983F
:10001000000005D004D036950A95B1F708951EE08A
:100020001A95F1F7089528EC18EC1A95F1F72A952E
:10003000D9F70895C895302D303019F0E2DFE395F7
:10004000F9CF089568656C6C6F20776F726C640DE2
:100050000A00C39ABB9AE4E4F0E0ECDFE4DFFBCFF4
:00000001FF
```

These are "hex" codes, short for hexadecimal, which is a numbering scheme that uses letters as well as numbers to count up to sixteen rather than ten for decimal digits. Base sixteen is commonly used for computer codes because the number 16 corresponds to four binary bits, which is a convenient size for a digital device to handle.

In a desktop computer, programs are stored on a rotating hard disk; in a tiny microcontroller with no moving parts they're saved in "flash" memory. This is a kind of electronic memory that retains information that's been stored even after the power is turned off, by holding onto the electrons that physically represent a bit. Such memories originally required special programmers to load information into them, but like so many other functions, those circuits have also been squeezed into microcontrollers so that any computer can generate the required control signals to write to flash memory. Here, the hello-world program is loaded into the circuit through a clip connected to a PC's parallel printer port:

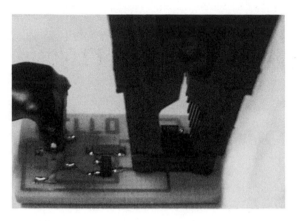

This process is naturally enough called in-circuit programming.

Finally, when the hello-world circuit is plugged into a PC's serial communication port, it turns on and starts endlessly sending back its message:

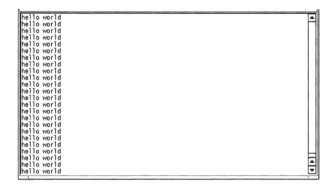

In coming chapters, we'll see how the world can say hello back.

Making Sense

Kalbag

S. S. Kalbag had an uncanny ability to make you feel better about the future of the human race, simply by walking into a room. In his quiet way he radiated an uncommon goodness that touched everyone who encountered him. I first met Kalbag at the Vigyan Ashram, the rural science school he ran outside a small village in the western part of India.

The name *Vigyan Ashram* is itself a gentle joke. "Vigyan" is an old Sanskrit word for all the subjects under the umbrella of the natural sciences. And "Ashram" comes from the ancient schooling system in India called *Gurukul.* Students would leave home to live in a secluded community of scholars, or ashram, to study with an enlightened teacher, the guru. So "Vigyan Ashram" is a retreat for the spiritual contemplation of the natural sciences rather than religion.

Kalbag grew up in what were then wilds outside of Mumbai. There were no services—if his family wanted electricity, or running water, or sewers, they had to provide these things for themselves. Which they did. He grew up not only learning how to build such systems but also

Kalbag

seeing engineering in the home as an opportunity rather than an obstacle. If he didn't like something, he fixed or improved it.

As Kalbag began his conventional schooling, he stood out as someone with great academic promise. He found it a bit odd, though, that he was scoring high marks but wasn't required to use his hands. He didn't understand why he wasn't being taught anything in school about the practical lessons he was learning at home.

Kalbag went on to earn a PhD in food technology, studying industrial food processing. After twenty-seven years as a student, he then planned to spend twenty-seven years in the "Grihasthashram" stage of Hindu life, that of a householder. He married his beloved wife, Mira, and began work in Mumbai for the giant consumer goods company Hindustan Lever. He rose through the ranks there, eventually becoming the head of the engineering division.

Always in the background was his interest in hands-on learning. In his spare time Kalbag did a study of street kids in Mumbai, wondering how they got there. Not surprisingly, he found that they dropped out of school because of the irrelevance for them of what was taught in school. But those same kids grew up to fix the cars and build the houses of the people who did go to school, using skills that their academically inclined employers lacked. They were clearly capable of mastering complex concepts that they cared about.

On schedule, and to the great consternation of his colleagues, Kalbag left Hindustan Lever in 1983 at age fifty-four to begin the "Vanaprasthashram" stage of traditional Hindu life, service to his community. His dream was to set up a working school, where students would learn science and technology through their relevance to everyday life. He wanted to do this where it was most needed, so he and Mira began touring the countryside. They found what they were looking for in Pabal, a tiny village a few hundred kilometers from Mumbai. It was poor, with little cash income, and dry, with poor water supply. But it was also rich in fresh air and open space. If he could make the school work there, he could make it work anywhere.

Kalbag's ultimate goal was to show that rural understanding and application of science was the key to reversing what was perhaps India's most pressing problem, urbanization. Because of a perceived lack of opportunity in the countryside, the cities are unsustainably crowded with villagers moving in, taxing support services beyond their carrying capacity. His dream was to reverse that migration by making rural life not just sustainable but desirable.

He built a school, based on the principle of learning-through-doing that he had first encountered as a child. Kalbag built the school literally as well as metaphorically; he and his students erected the buildings, down to the 240-square-foot cabin where he lived until his untimely death at seventy-four.

Vigyan Ashram took the dropouts from the educational system, the kids who were considered uneducable. The school wasn't a charity, though; his students paid to attend. But that was an easy sell because they made a profit on their investment in education, through the businesses they ran at and after school. Economic as well as environmental sustainability was always a key concern there.

For example, the ground resistance meters that were used for locating water in the area cost 25,000 rupees (about $500). At Vigyan Ashram they bought one, stripped it apart, and from studying it figured out how to make them for just 5,000 rupees. Those meters could then be operated to earn 100,000 rupees per year in a business. Another example arose because they needed a tractor on the farm at Vigyan Ashram, but could not afford to buy a new one. Instead, they developed their own "MechBull" made out of spare jeep parts for 60,000 rupees ($1,200). This proved to be so popular that a Vigyan Ashram alum built a business making and selling these tractors.

By the time I got to Pabal in 2001 it was clearly an innovative place. Instead of the crumbling dwellings seen in nearby villages, Vigyan Ashram was dotted with neat ferro-cement geodesic domes, designed to make efficient use of the available building materials.

In one of these domes, I was pleasantly startled to find in this remote setting a complete little chemistry laboratory. It was there for performing simple tests, such as checking blood samples for diabetes and pregnancy, or water samples for impurities. A group of women at the school was running a business out of the lab. Chemical testing was an opportunity for them because this was not a job traditionally done by men (since it was not a job traditionally done by anyone).

When I first met Kalbag, I casually mentioned the kinds of personal fabrication tools we were developing and using at MIT. In a not-so-quiet way he did the verbal equivalent of grabbing me by my lapels (he didn't actually grab me by the lapels was because I was wearing a T-shirt). He reeled off a long list of things they wished they could measure, but weren't able to.

Their soil meter could test for moisture, but not nutrients. In agribusiness, *precision agriculture* refers to the use of such agricultural

The MechBull

Vigyan Ashram

Better living through chemistry

Growing graphs

measurements to precisely match irrigation and fertilization to a crop's requirements, in order to minimize consumption and maximize production. In a rural village it is even more essential—what's at stake is not just a corporate profit-and-loss statement, it's the survival of the farm and the farmer. The need for environmental data is particularly acute in the face of changes in crops, land use, and climate at Pabal and around the world. Traditional practices aren't a helpful guide where there is no tradition.

At Vigyan Ashram one of the new crops they're growing is castor, which produces seeds that can be pressed to provide the most important ingredient for the MechBull, its fuel. However, optimizing an engine to burn castor seed oil requires advanced instrumentation to measure the combustion products and to evaluate the overall performance compared to conventional fuels. In fact, just tuning an engine requires instrumentation not easily available in rural India. The devices used at your corner garage cost more than a villager can afford, there isn't a supply chain to bring them there, and even if they were locally available the commercial products are intended to work in a garage rather than a rural field.

Kalbag's refrain across these examples was the need to quantify processes on the farm in order to understand and improve them. He had his students plot whatever they cared about against whatever might be relevant, so that they could work out for themselves answers to questions such as the right amount of grain to feed a growing chick, or the irrigation needed by different crops. The problem with this enterprise was that they could quantify only what they could measure, which excluded many of the things they were most interested in knowing. In part this was due to technical limitations on the available sensors. And in part it was a social and economic limitation, due to the vested interests in preserving that ignorance.

Take milk. It's essential for childhood development, but in the communities around Vigyan Ashram some distributors were diluting it with water and white powders, reducing the nutritional value. Even worse, good milk was being mixed with bad milk in the hopes that it would average out as OK milk. And dairy farmers were paid based on the milk's fat content, but that wasn't measured fairly or reliably. No single test could cover all of these attributes; near the top of Kalbag's wish list was the need for an instrument that could completely characterize milk.

Or take rice. This staple for both diet and the economy is bought from farmers in large regional markets. The price is based on the quality, weight, and moisture content of the rice, which is determined by the judgment of an employee of the rice market. I was joined on a visit to one of these markets by the local district collector (DC). That position is a legacy of the British Raj, during which the DC represented the Crown to collect taxes and dispense justice (from the name, guess which of those functions was the priority). Although DCs were and still can be notoriously corrupt, this man was honest. As the boss of the rice market explained to me how it operated, the DC leaned over and told me how the market was screwing the farmers left and right. When pressed, the market boss showed me a crude meter at the market that was supposed to impartially measure rice quality but—surprise—it wasn't working. The farmers wanted and needed an instrument under their control.

The recurring rural demand for analytical instrumentation led to an easy decision to launch a fab lab with Kalbag at Vigyan Ashram, to be able to locally develop and produce solutions to these local problems. The opportunity in Pabal was clear; less expected was the impact of the lab outside of the region.

In New Delhi, the head of the national electricity utility explained to me that about a third of the power they generated was stolen and lost in the distribution network. A common sight in an Indian city is a tangle of unauthorized power taps. Contrary to common perception, India could meet its present energy demand if this power could reach its intended recipients. The problem isn't enforcement; it's instrumentation. Because of the cost and complexity of monitoring power con-

Measuring rice

The grid

sumption, the utility wasn't able to map the reality of what was happening in the grid with enough resolution to spot where power was disappearing.

That discussion led to a sketch for an inexpensive networked power meter that could be widely deployed in order to catch power theft as well as reward efficient consumption. When I suggested that we then go try out the idea, they looked at me blankly: the grid's engineering office was designed to supervise billion-dollar infrastructural investments, but not invent anything itself. An engineer was duly dispatched from Delhi to a place in India that did have the tools for this kind of advanced rapid prototyping: the fab lab on Kalbag's rural farm.

A few months later, much the same thing happened in an even more surprising way. I had returned to the United States and was visiting the research labs of an integrated circuit manufacturer that was developing devices based on work at MIT. Those chips were designed by teams of engineers and produced in billion-dollar chip fabs. We were discussing new applications for the first version of the chips, and measurement strategies to use in the next generation. When I suggested that we try these ideas out, the reaction was the same as in Delhi: they looked at me blankly. There was no way for an individual with an idea there to invent something alone; the corporate development process demanded the skills of groups of specialized engineers, using expensive facilities distributed over multiple sites. When I jokingly suggested that we go to Kalbag's farm, the silence in the room suggested that the joke was very serious. The possibility that a rural Indian village had technical capabilities missing in a facility at the apex of the technological food chain had, and has, unsettling implications for survival at the top of the food chain.

I don't think that these stories represent either notable failings of integrated circuit and power engineers or unique attributes of Kalbag, remarkable as he was. Instead, they reflect the importance of necessity in invention. Unlike a farmer in India, the survival of a chip designer does not directly depend on the price of rice or the quality of milk.

Because Kalbag and his students had to produce both the food and technology that they consumed, the impact of access to tools to develop instruments was much more immediate for them than for an engineer surrounded by support systems. In retrospect it's not surprising that those tools should appear first where they're most needed.

Information technology requires information, and in Kalbag's world the most important information is about the state of that world. IT alone is useless at Vigyan Ashram without the instrumentation technology to interface between bits and bulls. Appropriate technology in this case is advanced technology. The Kalbags and Vigyan Ashrams of the world are leading rather than lagging behind more developed parts of the world in mastering the tools and skills for personal fabrication in order to make locally meaningful measurements.

Instrumentation

Kalbag's agricultural questions revolved around his need for instruments to quantify qualities of interest. The good news, and bad news, is that there are nearly as many sensors as sensees. It's possible to measure almost anything. The art as well as science of instrumentation is to do so in a way that is fast, cheap, and accurate enough.

Most measurements start with an analog (continuous) electrical signal. Common sensors include resistors that vary with temperature (thermistors), diodes that change an electrical current based on light intensity (photodiodes), and capacitors that store an amount of charge that depends on acceleration (accelerometers). In an ideal world, that would be all that these devices responded to; in our nonideal world, attributes that aren't of interest, such as the temperature dependence of a photodiode, can influence a measurement and so more than one sensor might be needed for calibration.

Analog signals enter into digital circuits through analog-to-digital converters, invariably abbreviated as A/D. These operate right at the boundary between analog and digital, with analog amplifiers and filters on one side and digital logic on the other. In the early days of computing an A/D converter was a whole instrument. Now they're available as separate chips, and as a circuit integrated into a multitalented microcontroller.

Sensors are ubiquitous, as close as the nearest person shouting into a cell phone. A microphone is a sensor that produces a voltage proportional to the sound pressure impinging on it. A microphone can be connected to the hello-world circuit:

and sampled by an A/D converter in the microcontroller. Now, instead of the circuit sending the characters "hello world" to the computer that it is connected to, the microphone can convert the sound waves of the spoken words "hello world" to a varying voltage, the A/D can turn the voltage into numbers, and the microcontroller can then communicate those numbers as a digital recording. This is what a graph of those numbers looks like:

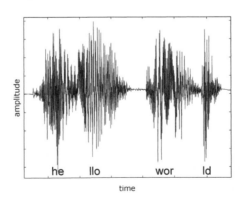

In "Interaction" we'll see how to play back this recording.

A/D converters are characterized by the rate at which they can convert a signal and the resolution of that conversion. Beyond an A/D in a garden-variety microcontroller that can sample an audio signal thousands of times a second, there are high-speed A/Ds with conversion rates that range up to billions of samples per second. At that speed they can directly digitize radio signals, a capability that is used in digital "software" radios. The available resolution of A/Ds ranges up to tens of bits; a twenty-four-bit A/D can resolve voltage changes as tiny as one part in 16,777,216. That's such a small fraction that it can reveal the thermal jostling of the electrons in a circuit, which sets a practical limit on the sensitivity of a measurement. A/Ds with that kind of resolution are used in demanding applications such as the radio receivers used by cell phone base stations (and eavesdroppers) to pick out the weak signal of a wireless handset from among the clutter of background radio-frequency noise.

Because the performance of high-speed and high-resolution A/Ds comes at a severe price in dollars and power consumption, A/Ds are usually chosen to be just good enough for a sensor and an application. There's no point in resolving the electrical thermal noise coming from a microphone if the background acoustic noise will be many times larger than that. Most common are eight- to twelve-bit converters (resolving one part in 256 to 4,096, respectively).

The most accurate A/Ds paradoxically use just a single bit, and even more paradoxically they intentionally add carefully calibrated noise to the signal. These one-bit A/Ds have a circuit called a comparator that outputs a 1 if the input voltage is above a threshold, and outputs a 0 if the voltage is below the threshold. This comparison can be done very quickly and very accurately. Because of the added noise, even if the input is steady the comparator output will fluctuate between 0 and 1. The relative fraction of time that the output is a 0 or 1 will change as the input voltage changes, providing a measurement of the voltage. That fraction can easily be determined by a simple digital counter, and the resolution of the measurement can be increased by counting longer.

A one-bit A/D is an example of a powerful measurement technique, converting a quantity of interest (in this case, voltage) into one that is a function of time (in this case, the fraction of time that the comparator output is 0 or 1). That's a good trade-off because time is one of the easiest things of all to measure. The one-dollar microcontroller used in the hello-world circuit can perform operations faster than a million cycles a second. This means that after just a few seconds of counting events from a digital sensor, such as the output of a comparator, it can resolve differences as finely as an expensive twenty-four-bit A/D. And, unlike an A/D with a fixed resolution, if the measurement is made by counting then it's possible to quickly get a low-resolution result and count longer to add resolution.

Unlike a microphone in the hands of a late-night lounge singer, a good sensor should not influence the behavior of the system under study. It's ideally an uninvolved observer. However, by actively stimulating a system it's possible to go beyond a description of what the system is doing and learn something about how it works. Active measurements rather than passive sensing can also aid in separating a desired response from other influences on the system.

As an example of this, here's another hello-world circuit, this time adding a resistor and a connector to apply a voltage to a sample and measure the transient response to the voltage:

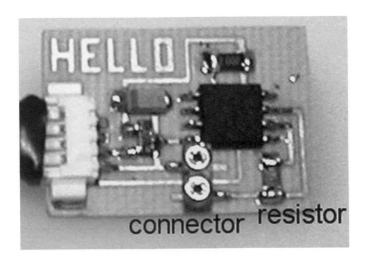

When the microcontroller applies a voltage to the output pin, current will flow through the connector and into whatever it is connected to. By measuring the rate at which current flows through the resistor, the microcontroller can learn about the material from its response.

For example, return to Kalbag's original question about a simple way to characterize the dilution and fat content of milk. A vinyl cutter can be used to cut out electrodes and a cable to connect this circuit to a glass of milk:

When the microprocessor applies a voltage to the electrodes, an electric field is created between them. This electric field exerts a force on the molecules in the milk, reorienting them and effectively changing the field strength. The more milk there is, the more charge the microprocessor must supply before the electrodes reach their final voltage. This relationship between charging rate and milk volume provides a precise way to determine the amount of milk in the glass.

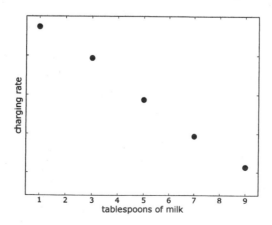

This offers an exact measurement of whether the glass is greater than half full, less than or equal to half empty, or the milk has spilled and crying is in order.

There are many ways that different materials respond to electric fields: They can conduct the field, internally align with or against it, or store it. And these things don't happen instantaneously; the rates depend on the details of the composition. This means that the time dependence of the response to an electric field provides information about not just the quantity of material but also its characteristics. Here's what the charging curves look like for different types of milk, as measured by the hello-world circuit after applying a voltage:

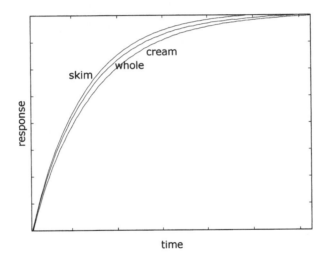

In each case it's the same amount of milk; the differences in the curves come from the details of their constituents.

The points in these curves are spaced in time by the fast clock cycle of the microcontroller, rather than the much slower rate at which the A/D can actually do conversions. That's possible thanks to the trick of "undersampling." Inside the microcontroller, there's a sample-and-hold (S/H) amplifier at the input of the A/D. The S/H circuit can quickly capture an analog copy of a rapidly varying signal, holding it steady while the A/D converts it to a digital value. The milk measurement is a response to the application of the charging voltage by the microcontroller. This means that the microcontroller can trigger the S/H to sample the response at a delay from when the charging starts. By the time the A/D has done the conversion the response to the charging will have long since settled, but the whole process can be repeated again with a slightly longer delay between the start and when the S/H grabs its sample. In this manner a series of measurements taken at different times can be assembled into a single curve as if it was recorded by a much faster (and more expensive and power-hungry) A/D. This is called *undersampling* because any one measurement gets less than the full signal.

The milk meter's electrical impulse response measurement can mathematically be transformed to plot the same information as a function of frequency rather than time. That's called a *frequency response measurement*. Before the advent of high-speed, low-cost microcontrollers, it was measured using much more expensive laboratory equipment to sweep the frequency of a signal generator and record the response at different frequencies.

Frequency response measurement is an example of a spectroscopy, meaning that something is being measured as a function of something else. Other important spectroscopies record the response of a material to different colors of light, or to high-frequency magnetic fields. Such curves or functions are usually much more informative than a simple number from a sensor, such as just the light intensity in the case of the electrical impulse response providing insight into the quality as well as quantity of milk.

Compared to analyzing milk, measuring the presence of a person is like falling off an electronic log for this example sensor circuit. The materials in people, like those in milk, respond to electric fields. Depending on the geometry of the electrodes, it's possible to determine the distance, position, or orientation of a nearby part of a person. The final hello-world circuit in this chapter replaces the milk-measuring connector with an electrode to serve as a button:

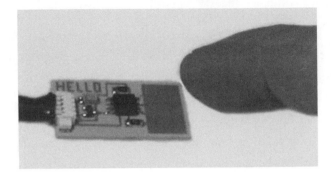

Here is how the charging rate of the electrode varies over time as a finger approaches and then touches it:

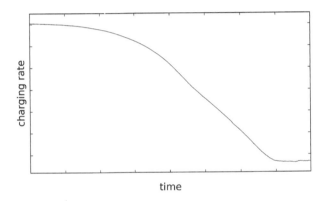

The button can detect the finger even before it's in contact. This kind of measurement provides an easy way to attach a user interface to an object, defining buttons, keypads, keyboards, sliders, and pointing devices all by the shape of underlying electrodes.

Net Work

If the practice of personal fabrication expresses a deep-seated human desire to create, a driving inspiration for that creation is communication. Around the world, one of the first things people do when they get access to tools for technological development is to apply them to accessing and exchanging information. This leads to some unexpected innovators in telecommunications infrastructure.

Sharp

In rural India, even the most humble mud hut is likely to have a television set, powered by a rechargeable car battery if there isn't electricity available. Since there's too much of rural India to be reached by urban broadcasts, more often than not the programs arrive at the TV via cable distribution of satellite signals. Those cable systems were built and are run by local businesses across the country, often bringing a cable connection long before telephone or electrical service arrives.

While the Indian government was busy regulating its unreliable telephone service, grassroots entrepreneurs covered the countryside with cable TV networks. By the time the government got around to trying to tax them in the mid-1990s, there were about a hundred thousand cable operators (the estimates are unreliable, because operators have a strong financial incentive to not be counted by the government). By then, the technological cat was out of the bag and this network could not be centrally controlled.

A typical village cable system might have a hundred subscribers, who pay one hundred rupees (about two dollars) per month. Payment is prompt, because the "cable-wallahs" stop by each of their subscribers personally and rather persuasively make sure that they pay. Visiting one of these cable operators, I was intrigued by the technology that makes these systems possible and financially viable.

A handmade satellite antenna on his roof fed the village's cable network. Instead of a roomful of electronics, the head end of this cable network was just a shelf at the foot of his bed. A sensitive receiver there detects and interprets the weak signal from the satellite, then the signal is amplified and fed into the cable for distribution around the village. The heart of all this is the satellite receiver, which sells for a few hundred dollars in the United States. He reported that the cost of his was one thousand rupees, about twenty dollars.

Amazed that such a sophisticated instrument could cost so little, I studied it for clues as to how it was made. The receiver was labeled "Sharp," but it clearly didn't come from the Japanese consumer electronics company, and there was no other identifying information. The operator reported buying it at a market in the electronics district in Delhi, so that became my next stop.

Every imaginable electronic artifact is bought and sold from crowded stalls in a warren of alleys in this market. Not only that, there's a section devoted to selling all sorts of specialized electronic components for making things, and another for while-you-wait repairs. Down a side alley was the section for satellite receivers, and in one of these was the local Sharp vendor who was indeed selling them for one thousand

Home theater

Global village

Cable operator

Silicon alley

rupees. He in turn got them from the manufacturer, which he reported was located in Delhi's lumber-cutting district.

This tip led me to an unremarkable door discreetly labeled "Sharp" amidst open-air furniture workshops. It opened onto a room full of people, sitting at workbenches with modern electronics test equipment, surrounded by the same kind of Sharp satellite receivers that I first saw in the rural village.

This Sharp turned out to be an entirely independent domestic brand. They produced all of their own products, although not in that room—done there, it would cost too much. The assembly work was farmed out to homes in the community, where the parts were put together. Sharp operated like a farm market or grain elevator, paying a market-based per-piece price on what was brought in. The job of the Sharp employees was to test the final products.

The heart of the business was in a back room, where an engineer was busy taking apart last-generation video products from developed markets. Just as the students in my fab class would learn from their predecessors' designs and use them as the starting point for their own, this engineer was getting a hands-on education in satellite reception from the handiwork of unknown engineers elsewhere. He would reverse-engineer their designs to understand them, then redo the designs so that they could be made more simply and cheaply with locally available components and processes. And just as my students weren't guilty of plagiarism because of the value they added to the earlier projects, this engineer's inspiration by product designs that had long since become obsolete was not likely to be a concern to the original satellite-receiver manufacturers.

The engineer at the apex of the Sharp pyramid was good at his job, but also frustrated. Their business model started with existing product designs. The company saw a business opportunity to branch out from cable television to cable Internet access, but there weren't yet available obsolete cable modems using commodity parts that they could reverse-engineer. Because cable modems are so recent, they use highly

Quality control

integrated state-of-the-art components that can't be understood by external inspection, and that aren't amenable to assembly in a home. But there's no technological reason that data networks couldn't be produced in just this way, providing rural India with Internet access along with Bollywood soap operas.

After all, Sharp and its partners had assembled a kind of fractal, hierarchical, entrepreneurial tree that spanned India. From the engineer in the back room, to the kitchen workshops doing the assembly work, to the regional electronics vendors selling the products, to the village cable operators using them, they were producing and distributing surprisingly advanced electronic instruments out of sight of India's academic and industrial establishment.

With the narrowest of financial motives, this collective enterprise had succeeded in doing large-scale sustainable technology deployment. They weren't on the visiting agendas for aid organizations or foreign investors; it's hard to even know where to visit, and there are plenty of visitors they'd rather not see. There isn't even a single entity with which to partner on a joint venture; the whole operation is fundamentally distributed. But India's rural cable system shows how something as sophisticated as a communications network can be created by something as informal as a human network with an appropriate incentive.

Haakon

Haakon Karlsen is a farmer and herder in the Lyngen Alps, at the top of Norway far above the arctic circle. There, satellite TV dishes aim down at the horizon rather than up toward the sky, since that's where geosynchronous satellites orbit. But Haakon is less interested in hearing from satellites in space than from his animals in the mountains.

In traditional Sami herding, the sheep flocks spend the summer in the mountains, going up in May and coming back down in September. Their reindeer herds don't even come down, wintering in the mountains. Finding the animals when it's time to bring them back is a perennial problem, as is getting the right animal to the right herder. Then there are new problems such as predators appearing—wolves, bears, and lynx—because of the loss of traditional hunting grounds elsewhere. One in ten sheep don't survive the season; there's a great need to be able to recognize at a distance when they're in danger, injured, or sick.

The traditional technology for tracking sheep is a bell attached to the lead sheep of the flock. This works well if a herder follows the flock to be close enough to hear it. Haakon wondered why he could receive on his farm a radio signal from a satellite in outer space but not one from a sheep in the mountains, and he set out to correct that.

There wasn't an incumbent network operator to service the sheep in the Lyngen Alps, so, like an Indian cable wallah, Haakon had to build whatever telecommunications infrastructure he needed. Here, though, the customers had four legs instead of two.

Bell sheep

Network and user

He was helped by Bjørn Thorstensen and Tore Syversen from the Tromsø office of Telenor, Norway's national telecom operator. The team came up with a short-range sheep radio, similar in size to the ear tags worn by sheep as identification, but now sending the animal's ID electronically. A larger "bell" tag, worn by the lead sheep, contained a satellite GPS receiver that could report the location of the herd. Telenor initially put these together, then in 2003 a fab lab opened on Haakon's farm to continue development and production of the radios and antennas there.

The sheep radios had a range of a few hundred meters. More powerful radios would have been too large for the animals to wear comfortably, and the batteries wouldn't last through the season. The short range of the sheep radios required the installation of fixed radio repeaters to relay the signals down to the farm. These repeaters used cheap consumer radios from commercial wireless networks, arranged in a "mesh" configuration that could pass a signal from radio to radio in order to detour around mountains.

It was a bit adventurous for Telenor to help with Haakon's experiment, because mesh wireless networks present a serious threat to its traditional business model. Instead of hopping a signal from pasture to pasture, data could also pass from house to house or town to town. Each user could add an antenna to join the network, and in so doing simultaneously extend the network by relaying messages for others as well as sending and receiving their own. If end users could build their own networks, why would anyone pay to use Telenor's?

Telenor's involvement was also wise. A farm above the arctic circle is a safe place to explore the explosive commercial implications of this kind of grassroots networking; it may be possible for community networks to comfortably coexist with larger-scale commercial ones. And, who knows, four-legged customers might represent an untapped new telecommunications market. After all, cell phones have already been sold to just about every available two-legged creature in a communications-crazy place like Norway.

Vicente

Around the turn of the twentieth century the architect Antonio Gaudi pioneered a fluid, expressive building style in Barcelona. His improbable, whimsical, organic shapes defy convention and even apparently gravity. Looking at his Temple of the Sagrada Familia, started in 1882 (with work continuing up to this day), it's clear that he didn't subcontract his civil engineering. The structural elements are gracefully integrated into the visual and social design of the structure. To accomplish this he in fact did make innovative contributions to engineering practice. For example, he designed his arches in upside-down models by hanging weights on a chain to represent the loads on a building and then using the resulting curves as templates to position structural supports.

Civil engineering

Vicente Guallart comes from a new generation of architects in Barcelona, picking up where Gaudi left off to mold the materials of our day: computers and communications. He's depressed by the lack of connection between the design of the shape of a building and the design of the information technology within the building. One team of designers and builders makes the physical structure, and then when they're done another group comes along and tosses some computers

Media House

over the transom. Vicente wondered if the electrical engineering in a building could be as expressive and integrated as Gaudi's structural engineering.

Vicente's question extended an earlier series of collaborative explorations of embedded intelligence that we had done at MIT, including an installation of interactive furniture in the Museum of Modern Art in New York City (described in "Art and Artillery"), a communicating bathroom shelf that could help seniors manage their medication (shown in a demonstration of future technologies at the White House/Smithsonian Millennium celebration), and a smart stage for the juggling Flying Karamazov Brothers that let them control sights and sounds through their motions. A natural next step was for us to work together to build a building.

The result of this collaboration was a test structure called the Media House, presented in a theatrical event in Barcelona in 2001. The vision

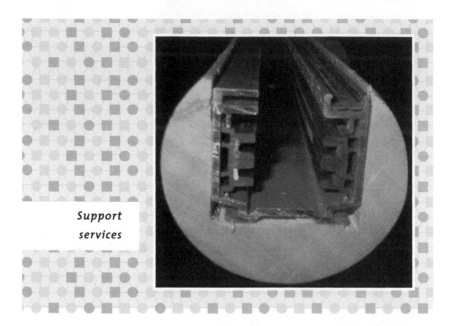

was to integrate into the design not just mechanical support but also all of the services, so that the structural elements could carry energy and information as well as physical forces. The physical construction process should simultaneously erect the building's functional infrastructure.

In the construction of buildings today, detailed drawings show in advance the location of every wire, switch, fixture, and outlet. These plans must be laboriously drawn, followed, and checked, and the infrastructure remains fixed until it gets ripped out in a remodeling. The wishes of the occupants must pass through architects, engineers, builders, contractors, subcontractors, and installers, a process that is as (un)reliable in construction as it is in a party game.

The Media House was designed so that nothing in the infrastructure was fixed. It was built out of movable beams containing tracks carrying power and data. Individual elements such as lights and switches fit into these tracks. Each component contained a tiny computer that understood the Internet's communication protocol, so that a fixed wiring diagram wouldn't restrict their connections. The operation of lights

Internet site

and switches, thermostats and heaters, buttons and bells could be programmed based on their patterns of use as well as the demands of architectural expression and energy efficiency.

All of this was possible with just a few dollars in added parts for each device. Internet implementations on desktop computers are a kind of technological equivalent of human bureaucracy; much of the computer code spends its time passing messages between software layers developed by different groups of people, effectively reading and writing memos instead of doing useful work. That software stack can be significantly simplified for the limited and predictable communications needs of a lightbulb.

Even if it is possible to connect a lightbulb to the Internet at a cost compatible with a disposable device, it would be pointless if the lightbulb demanded the services of a corporate IT department to manage it. So, in the Media House each of the elements of the infrastructure also stored the data on its associations, and the procedures for updating them, so that assembling the pieces simultaneously built a network, a database of information about the configuration of the infrastructure, and a computer to process that information. The occupants of the house rather than its builders put this together, so it can grow and change to reflect their needs. And there's no central computer to configure and manage and that would be a vulnerable central point of failure.

At the opening of the Media House, one of the computer architects of the high-speed "Internet 2" project kept asking how fast data could be sent through the structure. I reminded him that lightbulbs don't need to watch broadband movies, and that in fact a network must be much more complex to send data that fast. That complexity makes the network more expensive to build and difficult to change; installation and operational costs are the key questions to ask about a networked lightbulb, not communications speed. He persisted in asking about the data rate, until in exasperation I joked that the Media House was not part of the new Internet 2, it was part of an "Internet 0" (counting today's network as Internet 1).

The joke stuck and came to refer to a set of principles that bring Internet connectivity to everyday devices. If a computer scientist had trouble seeing the value of slowing networks down to simplify them, the building construction industry didn't. In the United States it's a trillion-dollar-a-year business, dwarfing computing and communications. There's tremendous demand in the industry for bringing programmability to buildings so that it's not necessary to laboriously draw the wiring diagrams, have installers follow them and inspectors check them, and then contractors redo them if the occupancy of the building changes. Instead, the configuration of Internet-connected lightbulbs and thermostats could be controlled by the building's occupants, with the builder's job reduced to ensuring that the devices get power and information.

The question about the data rate to a lightbulb reflects a long-standing bias in the research community that ever-faster networks and computers are better than slower ones. In the life of a technology, there is an early "talking dog" phase: it would be notable if a dog could talk at all; what the dog first says wouldn't matter as much. Only later do you begin to care what the dog talks about. Likewise, in the early days of networking, it was impressive enough just that the networks could work. It didn't require much insight for a researcher to recognize that a network would be more useful if it was faster. But we've long since left the talking-dog phase of networking and its easy recipe for progress through increasing performance. Fewer and fewer applications are now limited by the available higher and higher speeds. Instead, the technological challenge faced by the building industry is not one of absolute performance; it's the cost of complexity. Their need is for computationally enhanced materials that behave like any other construction material, so that the skills of a building architect, network architect, and computer architect are merged into the fundamental act of assembly.

In their own ways, an architect in Barcelona, a herder in the north of Norway, and a cable operator in rural India are all connected by their roles in creating communications infrastructure. The needs of each were not being met by incumbent operators in their respective worlds, leading each to take on a hands-on role in developing as well as using systems to solve their problems. And each example illustrates an ingredient in not just personal but community-scale fabrication of those systems: grassroots electronics production and distribution, mesh networks assembled from commodity components, and the merging of computation and communications into functional materials. At that intersection can be found the technological tools that are appropriate for the grace of Gaudi, the drama of Bollywood, and the splendor of the Lyngen Alps.

Communication

Communication between machines, like that between people, requires some common understanding and some civility, and frequently entails some compromise.

The first step in adding a communication capability to a personal fabrication project is to decide how information will be represented. The standard choice for turning human-readable characters into machine-readable numbers is ASCII, the American Standard Code for Information Interchange. This is a code that uses eight bits to represent a letter of the alphabet. It provides a table that a computer can use to convert between the characters that people understand and the numbers that computers use internally; Eight 0s and 1s correspond to 255 possible values. Beyond the alphabet, the extra entries are used for accented characters and control characters such as tabs and carriage returns. Written in ASCII, this is what "Hello" looks like:

01001000	01100101	01101100	01101100	01101111
H	e	l	l	o

The specification of ASCII as a standard settled around 1968. There were earlier standards, such as IBM's EBCDIC (Extended Binary Coded Decimal Interchange Code) that descended from punched cards and was used on mainframes. And there are later standards, most important, the universal character set (UCS) of Unicode. The "American" part of ASCII reflects a kind of techno-centrism, dating from when the computer industry was primarily an American enterprise. Unicode uses the same codes as ASCII for Roman characters, but uses sixteen or more bits rather than ASCII's eight to represent other alphabets from around the world.

ASCII or Unicode serves as a dictionary to translate a text message into a string of binary bits. Once that's done, the next step in digital communication is to choose how to physically represent those bits. The hello-world circuit in "Computation" used the venerable RS-232 spec (recommended standard number 232 from the Electronic Industry Association). This standard settled around 1969, originally to fill the need for connecting terminals to mainframes through links such as a telephone modem. RS-232 is used for serial communications, meaning that bits are sent in sequence down a wire, as opposed to a parallel link that uses many different wires at the same time. Parallel connections can be faster, at the expense of requiring extra wires.

The original RS-232 spec defines −12 volts to correspond to a 1 bit, and +12 volts to be a 0 bit. Eight ASCII bits, called a byte, are sent as these high and low levels, preceded by an extra 0 as a start bit to mark the beginning, and ending with a 1 as a stop bit. Here's the RS-232 signal for the *H* in *Hello:*

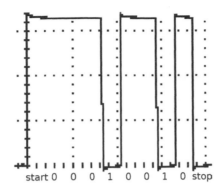

start 0   0   0   1   0   0   1   0 stop

Note that in RS-232 the bits in a number are sent out in order from the least to the most significant, so that in this signal they appear reversed from how they are written and stored (01001000).

In this example, the hello-world circuit is actually taking a common technical shortcut. Unlike a big computer, it's inconvenient for a little device that might be operating on 3 volts (as this one is) to generate ±12 volts. Officially, the range between –3 and 3 volts was defined as a "dead zone" intended to cover any noise in a communication line, but many of the products of today's electronics industry save power by operating at lower voltages that fall in this dead zone. Fortunately, the circuits that receive RS-232 signals increasingly accept the spirit as well as the letter of the standard. As long as the signal has the right shape, it's OK to use a lower voltage.

USB, the universal serial bus, succeeded RS-232 in 1995. USB can carry RS-232, but it can also run much faster, and it provides a way for standard kinds of devices such as keyboards and mice to identify themselves to the computers they are connected to. Also, instead of illicitly stealing power from one of the RS-232 communication pins, as the hello-world circuit does, USB explicitly provides power. But USB is still a serial point-to-point connection, meaning that like RS-232 it links whatever is at opposite ends of a wire. Electronic monogamy has its limits, though; the value of most any kind of device generally goes up with the number of things that it can talk to. Beyond individual links lies the world of networks.

The biggest network of all is of course the Internet. The Internet is a network of networks, globally connecting national networks made up of regional networks, which in turn connect networks spanning cities, buildings, offices, and homes, all the way down to your desktop. Internet connectivity was once the domain of mainframes, but it can now be embedded into a personal fabrication project. To equip something to talk to almost anything, it should be able to communicate via the Internet protocol (IP).

IP is the electronic equivalent of an envelope. A chunk of data to be sent over the Internet, such as an e-mail message or a Web page, gets assembled in an IP "packet," with the information at the front of the packet (called the header) containing the addresses of the sending and receiving computers. At the junctions where the Internet's networks meet, there are special-purpose computers called routers that read the addressing information in a packet and choose the best route for the packet to reach its destination.

This is called *packet switching*. Unlike an old-fashioned telephone switch that dedicates a wire to each call, in the Internet one wire can carry a mix of packets going to many different places. And the routes aren't limited by how the wires are connected; the routers can change the path a packet takes to its destination based on demand, capacity, and disturbances in the network.

The "Inter" part of the Internet comes from the concept of "internetworking." Before the Internet came along, there were many different kinds of incompatible networks. Messages couldn't pass between ethernet and token-ring circuits, or from the ARPANET to the SATNET to the PRNET systems. After the Internet arrived, there were still many different kinds of incompatible networks. But IP served to hide those differences from the computers connected to the networks—as long as they understood the IP protocol, they could all talk to one another.

IP's job is to get a packet from a source to a destination. If it acts like an envelope for data, then other standards specify the format of the contents of the envelope. These are conventionally described as a stack of software layers above IP handling higher-level tasks. First comes a layer for controlling communications. The simplest option for doing that is UDP, the user datagram protocol. This simply adds "port" numbers to the packet. Much as a ship coming to the United States might unload containers in Long Beach or oil in Houston, Internet port numbers serve as software destinations for packets arriving at a computer to reach the appropriate software applications. One port is for e-mail being sent to the computer, another for file transfers, and so forth.

An alternative to UDP for controlling communications is TCP, the transmission control protocol. This is like registered mail: TCP checks to make sure that a packet actually reaches its destination. Also, if a message is too large to fit into a single packet, TCP chops it up, sends the fragments, and then reassembles them at the other end. All of this requires extra communication, so TCP is used for applications where reliability is more important than speed.

Still higher layers handle the content of applications, such as the hypertext transfer protocol (HTTP) used by the World Wide Web. This standardizes the way pages of text and graphics are sent in an IP packet. "Hyper" isn't a value judgment about the frenetic nature of so much of what's available on the Web; it refers to enhancing the text on a Web page by embedding links to other pages or media types.

The new applications that have defined the Internet, from e-mail to e-commerce to e-everything-else, were introduced without having to reprogram the computers that operate the Internet, because these applications all still use IP packets. This approach has come to be called the "end-to-end" principle: What the Internet does is determined by the things connected to it, rather than being fixed by the design of the network itself. It's much easier for a programmer to develop a new Internet service and a user to install it as an application on a PC than it is for everyone involved in running the Internet to agree on reprogramming the routers and associated computers that make it work.

A defining image of the Internet was a cartoon of a dog typing on a computer, telling another dog that the great thing about using the Internet is that no one needs to know that you're a dog. But the dog still needed a conventional computer; the cartoon could now be updated from pets to a light switch talking to a lightbulb over the Internet. As long as a device can send and receive an IP packet, it can communicate with anything else on the Internet. This originally meant mainframes, but now even a simple microcontroller can do it.

The most complex part of Internet access hasn't been implementing the IP protocol; it's been the engineering details of actually connecting a device to the network. In the layered description of the Internet, below IP there are layers defining how a packet is physically represented and carried. An IP packet could be sent via an ethernet cable, or a Bluetooth radio, or a Wi-Fi hotspot, or a dial-up telephone modem, or a mobile phone, or a satellite terminal. Each of these transports has required dedicated chips to generate and interpret their signals, with multiple sets of chips needed for a product to be able to use more than one kind of network.

Internet 0 (I0) is an emerging alternative that simplifies and unifies the physical side of Internet access by using something like a Morse code for the Internet. Morse code communicates through the timing of impulses; it could be clicked on a telegraph, or flashed with a light from ship to shore, or banged out on a pipe. The signals are all the same; they're just carried by different media. The idea that grew out of the Barcelona Media House testbed described in "Network" is to do the same for a packet.

I0 sends an IP packet serially as ASCII bytes, like RS-232, but instead of defining voltages for a 0 and a 1, it defines times. Each bit is divided into two subintervals. If an impulse occurs in the first half, the bit is a 1; if the impulse occurs in the second half, the bit is a 0. The boundaries of the bits and the bytes are determined by putting impulses in both intervals for the start and stop bits that come before and after a byte. The spacing between these framing impulses also serves to show the rate at which the byte is being sent. This is what *H* looks like, now clicked in I0 impulses:

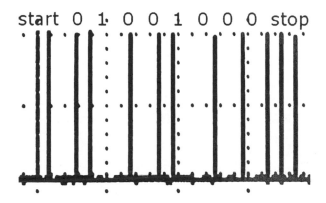

These impulses were generated by a modification of the hello-world circuit:

All of the other hello-worlds have been tied to a single computer; this one can connect to a network of them. It adds two extra connectors to distribute power to devices on that network. Those power lines can simultaneously carry I0 communication impulses as well as power, coupled onto and off of them through a capacitor. The voltage regulators in the circuits that they connect to won't care about those impulses—

whether the supply is impulsive or steady, it's all just available power for the regulator to filter. But the processors that are being powered can use the power line to distribute information as well as energy. The serial connection on the board carries messages from the power line network to and from the rest of the Internet. For extra credit, this hello-world circuit adds an LED so that it can flash when the world says hello to it.

Here's what a full Internet 0 packet looks like, and where the most important parts are:

This isn't just a picture of a packet; it *is* one. It looks a bit like a barcode and can be used that way. The raw signals from an optical scanner swept across it can go directly onto an Internet 0 network and be understood by any other device on that network. Internet 0 impulses can travel through anything that can carry a disturbance: They can be pulsed over a DC power line as the hello-world circuit does or over an AC power line, flashed optically by an LED or laser, silently clicked ultrasonically, radiated as a radio wave, printed on a page, or stamped into an object. Conversion among these different physical transports requires changing only the interface devices, not the signals sent over them. And communication with the existing Internet just requires sending the packet through its media access hardware.

There are twenty or so other standards aimed, like Internet 0, at networking embedded devices rather than general-purpose computers: CAN, LIN, BACnet, CEBus, LonWorks, X10, USB, Bluetooth, SPI, I^2C, ... These all share one feature: incompatibility. Dedicated converters need to be configured for any of these to talk to any other. This technological Tower of Babel was the result of a combination of historically distinct niches, proprietary development, and specialization for particular applications. Each of these standards does do something better than any of the rest. But in engineering more can be much less, because of the cost of optimization and specialization.

For any one kind of physical link, wired or wireless, acoustic or optical, there is an optimal way to represent a bit in order to send information as quickly as possible. This entails using a modulation scheme that takes advantages of the details of the link's amplitude response, time delays, and interference across the available frequencies. Using an optimal modulation allows bits to be packed together more closely than the time it takes for one of Internet 0's impulses to settle. That settling time depends on the speed of the signal over the size of the network, which is about a millionth of a second for a building-scale network. However, if there's no need to send data to a device in the building faster than a million bits per second, then a simpler option is to just wait for the impulse response to settle—hence the comment about lightbulbs not needing to watch movies, which first led to the name Internet 0 as a joking counterpart to the high-speed Internet 2. By slowing communications down to the rate it takes for a signal to settle over a particular network, the details of the network can be ignored and the same representation can be used across networks.

Internet 0's relationship to today's many incompatible communication schemes is a bit like the Internet's original relationship to the many incompatible networking schemes back then. The IP protocol provided a common language that could be shared across heterogeneous networks; I0 provides a common signal that can be shared across heterogeneous devices. Following this parallel with the origin of the Internet, I0's impulse encoding that's independent of the type of device carrying it has come to be called *end-to-end modulation*, and it enables "interdevice internetworking" among unlike devices. I0 trades optimality for simplicity and generality, which are as sorely needed in information technology as they are in the rest of the world.

Art and Artillery

Corporations sell to consumers as individuals, but individuals live in communities. And those communities can have compelling collective technological needs that are not met by individual acquisitions, notably in the ways groups of people can access and interact with information.

The unsuitability of computers that are designed for a person but need to be used by a group has led community leaders around the world to take on technological development projects to create computers that are appropriate for their world. In some of the most beautiful, and dangerous, places on the planet, these efforts are literally as well as figuratively on the front lines of a kind of fabrication that is not just personal, it's essentially social.

Terry

Terry Riley is the chief curator for architecture and design at New York's Museum of Modern Art. About once a decade MoMA mounts a landmark architectural exhibition that takes stock of the kinds of buildings being built. In 1998 Terry contacted me with a problem: For an upcoming show there was a great deal of new digital media to include, but he was looking for an alternative to the use of conventional computers in the galleries to access this information. MoMA goes to great effort to create a compelling visual and social space in the museum; he didn't want to destroy that with the solitary experience of hunching over a computer to type on its keyboard. Terry asked whether the furniture in the gallery could itself be used as an information interface.

Now, designing the details of computer interfaces is more typically the rather dry business of computer scientists and electrical engineers, but in this case Terry felt that those engineers weren't considering how groups of people rather than individuals could use computers, in a beautiful setting, without any instruction. So, with the show looming, he and his colleagues took it upon themselves to piece together a graceful, unobtrusive communal computer.

Lab space

We assembled a team of MoMA curators and MIT students to work on the project. The team put together prototype components at MIT, then we all decamped to midtown Manhattan to set up shop in the gallery for the final development. This was a striking, and challenging, site for a working lab.

The collaborative development process settled on a giant dining-room table as the focus of the exhibit. This had displays projected on it, and sensors in it to detect people as well electronic tags in objects. The goal was to retain the metaphor of architects looking at blueprints, but make the blueprints active. A central lazy Susan was dotted with what looked like coasters for drinks, each containing an image of a project in the show; as these tangible icons were brought near a place setting, the associated blueprint would magically appear (thanks to readers in the table detecting the tags in the coasters). Then the table would sense the viewers' hands, allowing museum visitors to browse through images and videos embedded in the plans. Particularly

Experiencing art

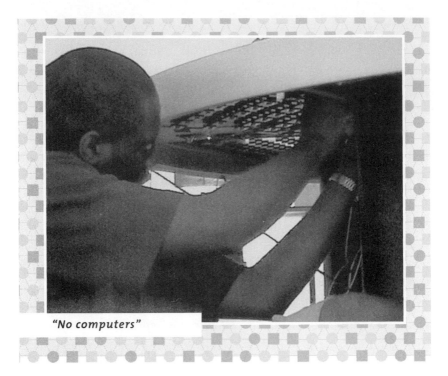

"No computers"

interesting pictures could be electronically slid onto the lazy Susan and shared around the table.

The designing and debugging continued right up to the opening of the show. As the minutes ticked by, we didn't realize that a few galleries away a wall of guards was holding back a sea of humanity. The guards parted, the sea flowed, we were swept away, and a crowd congregated around the table. We weren't sure how well, or even if, the table was working until an elderly museum benefactor came shuffling out of the crowd. Beaming, she said, "I hate computers, but this is great because there are no computers here!"

The museum benefactor didn't realize that the table she was pounding on housed seventeen Internet-connected embedded computers, communicating with hundreds of sensor microcontrollers covering the bottom surface of the table. But she was also right, in that there were no visible computers. By bringing that much processing power that close to people, the computing literally disappeared into the woodwork. Neither she nor I could tell exactly which device was doing what. The multiple tag readers, hand detectors, lazy Susan sensors, video projectors, and information databases were all interacting over the Internet. The table was not a peripheral to a computer; it was an essential part of MoMA's infrastructure for the show.

The table solved one problem—the intrusion of computers into the gallery—but caused another one: Some of the museumgoers were more interested in the interface than in the rest of the show. Although MoMA isn't (yet) in the business of technology development, visitor after visitor asked where they could buy an interactive table for use in their communal computing setting: teaching, or business management, or financial investing, or emergency services, or military command and control. The creative demands of exhibiting art had inspired the creation of a computer interface that was applicable far beyond a gallery.

Sugata

The most intriguing interest in the MoMA furniture came unexpectedly from Delhi's slums. There, Sugata Mitra was facing some of the same challenges Terry tackled in midtown Manhattan.

Sugata is a computer scientist. His office, like many in India, abuts a slum. There is a wall between the slum and his office, literally and figuratively. While idly musing on how to breach the barrier that wall represented, it occurred to him that he could just breach the wall itself. As an experiment, he opened a hole in the wall and stuck into it an Internet-connected computer facing the slum. There was no explanation, no instruction, just a computer monitor and a joystick.

Within minutes kids appeared, curious about this new arrival. They quickly worked out how to use the mouse to navigate, and from there they figured out how to surf the Internet. The kids in the slum then organized impromptu classes to pass on this knowledge to their friends.

A hole in the wall

And all of this was being done by children who were thought to not speak a word of English, the language used by the computer.

Because they were self-taught, the kids developed their own explanations for what they were doing. The mouse was called a *sui*, the Hindi word for needle, because it looked a bit like one. And the hourglass icon was called a *damaru*, after the shape of Shiva's drum. But in their own way, and on their own, they succeeded in learning to use the computer.

Sugata's jaw dropped when he came back to check on the computer and found on the screen the words "I love India"—he hadn't provided a keyboard, just the joystick! When Sugata asked the children how they were able to type the words, they explained that they had discovered the "character map" program on the control panel that can be used to enter arbitrary characters with the mouse by clicking on an on-screen keyboard. Sugata has a PhD, but he didn't know how to do that.

This experience was successfully repeated over and over across India. We all joke about how kids pick up computer skills better than their parents, but that observation had not been taken seriously as a pedagogical principle. Sugata termed it *minimally invasive education*. Rather than assuming that technology must arrive with armloads of explanations, and that everything must be translated into local languages, Sugata embraced the desire of a motivated student to learn. He asked instead how simple interventions could help his unconventional pupils along.

Sugata's biggest problem wasn't the slums, or the kids, or the computers; it was the interface devices. Mice and keyboards are intended for a single seated user, rather than a pack of kids clustered around a hole in the wall. And, even worse, those devices didn't survive in the slum because they weren't designed to stand up to monsoons, or sandstorms, or inquisitive animals. The interfaces started out as inappropriate, and ended up broken.

Terry Riley didn't have to contend with curious yaks in midtown Manhattan, but had his own kind of wildlife to worry about. The challenges of bringing computing into a museum and a slum are not really all that different. Both Terry and Sugata needed computer interfaces that could be used by groups rather than individuals, that could stand up to the enthusiasm of those groups, and that would fit into social situations rather than force antisocial behavior. Sugata, like Terry, was driven by the needs of his community to innovate in the design of computer interfaces beyond the limits of what was being done by engineers employed to develop those products for a living.

Sugata saw a remarkable consequence of getting that interface right. In the economic pecking order of an Indian city, there is a clear chasm between the haves and have-nots. Ownerships of cars, houses, televisions, even electricity and plumbing are status symbols of power and influence that divide kids within and beyond the slum. But the hole-in-the-wall computer was something different. Power with it came from knowledge rather than possession.

Haves and have-nots are adversaries; the have-nots generally want to get whatever it is that the haves have. But knowledge, unlike possessions, can be freely shared. Instead of haves and have-nots, around the hole-in-the-wall computer there were "knows" and "know-nots," kids who did and didn't know how to use it. It didn't help a "know-not" to attack a "know," however; the route to knowledge about how to use the computer was through friendship rather than conflict. Sugata's experiment in building stronger computer interfaces unexpectedly led to building stronger human interfaces.

Arjun

The social implications of prototyping computers for communities rather than for individuals led me from the intellectually rarefied air of a Manhattan art museum, through Delhi's slums, to the truly rarefied air at the top of the world in the Himalayas.

The Indian General Arjun Ray was responsible for protecting the restive state of Jammu & Kashmir, the Chinese border, and the

The front lines

Pakistani border. That is, he was in charge of the world's active nuclear battlefield.

When he arrived at his headquarters in Ladakh, General Ray had the assignment of ensuring border security. Conventionally, this meant the management of troops and weapons on the border. But he quickly came to the conclusion that border security follows from human security, human security follows from human development (intellectual, social, and financial), and the best way to invest in human development is through information. He saw that in the areas under his command there were so few educational, economic, or entertainment opportunities that there was little to do but shoot at one another.

This observation led him to launch "Operation Sadbhavana." *Sadbhavana* means peace and goodwill; General Ray sought to help connect these isolated communities with the rest of the world by providing Internet-connected computers that could be used for education,

Computer center

healthcare, communication, and commerce. By not lobbing a few extra cannon shells, he could afford to make these community investments.

Just as microbanking cooperatives have focused on women's groups, he particularly targeted the Buddhist and Muslim girls because they grew up to form the responsible backbones of their communities. When General Ray first arrived, the girls would run and hide from outsiders, who were seen as a threat to their community. The first success of the computer centers was to change that attitude; as the girls learned about them, they started to come running.

Visiting General Ray, I got caught in the worst snowstorm in the Himalayas in recent memory. This was fortunate, because during the storm, instead of just being an observer, I necessarily became an active participant in the project. While the winds blew outside, we set up one of the huts to function as a Web server rather than just a client, allowing the girls to create as well as browse Web pages. The response was electrifying. The girls posted pictures and descriptions to share with friends and family who might be just a few villages away, but it took many days' travel through the mountains to see them. It was eye-opening for them to think of themselves as sources rather than sinks of information.

I was in the Himalayas because, like Terry and Sugata, General Ray wanted computers that were designed to go into communities rather than just onto desks. Like Haakon and his herds in Norway, General Ray needed to deploy his own communication infrastructure because companies weren't competing to provide connectivity in the sparsely settled, inhospitable mountains. And, like Kalbag and his rural schools, General Ray wanted a way to produce the technology locally to meet the local needs, because cutting back on cannon shells is not a sustainable business model.

Not too long afterwards, I was in Washington, DC briefing a roomful of army generals. These were the best and the brightest of their generation, there to get guidance on the military implications of emerging technologies. After discussing personal fabrication and research on machines that make machines, I closed by telling the story of General Ray and his interest in local labs for network deployment as an investment in community stability.

Mental gear-grinding followed. The assembled generals understood General Ray's cost-benefit calculation, but they weren't sure how they could act on it. After all, there wasn't a Pentagon Office of Advanced Technology for Not Fighting a War.

Remote communities in unstable parts of the world may have technological needs surprisingly similar to those of the military, and out of necessity these communities can be even more advanced in adopting tools for personal fabrication rather than relying on long supply chains. But whose job is it to help those communities with their tools and technologies? The military, which is designed to break rather than build things? Aid agencies, which avoid any kind of speculative risk in investing in quantifiable social impact? Research agencies that support the investigation of speculative risks but avoid anything that looks like aid? Corporations, which develop products to sell to consumers rather than buy products developed by consumers?

The answer isn't clear. But that may in fact be the answer. These organizational distinctions have been based on a clear division of labor between the users and the developers of a technology, and may no longer make sense if almost anyone can make almost anything.

At one time a computer was a job description rather than a machine. Before electronic computing machines, *computer* referred to the teams of people who operated simple mechanical calculating machines to solve complex problems. These human computers were programmed by their written instructions. Likewise, typesetting was once a career for a trained typesetter, laying up lead type, rather than an expected menu choice in any word-processing program. It may well be that in the world of personal fabrication it is the fate of engineering to similarly become a shared skill rather than a specialized career.

From a New York museum to a Delhi slum to a front-line Himalayan village, the effort to adopt emerging technology to meet local community needs may migrate beyond the activities of visionary pioneers and become a common competence within and among those communities. Turning swords into plowshares might have been an unrealistic dream, but then again it was also an impossible one without widespread access to the tools to do so.

Interaction

Unlike an obedient child, most machines are meant to be heard as well as seen, or in some other way to alter their environment. This last of the tools chapters looks at some of the ways that things can communicate with people.

First, displays. The simplest display of all is a light, showing if something is on or off. Here's the hello-world circuit with a light-emitting diode (LED) added:

resistor LED

In addition to the LED, the other component added to the board is a resistor that limits the current flowing through the diode (unless of course one's goal is to explode an LED and rather dramatically launch it across the room, as many a would-be electronics hobbyist has discovered by accident).

LEDs are used instead of lightbulbs, here and almost everywhere, because they're smaller, more efficient, last longer, and they have also become cheaper. Because an incandescent lightbulb makes light by heating a filament, most of the energy is wasted as heat. An LED emits light without heat by doing the equivalent of rolling an electron down a hill; when the electron gets to the bottom the excess energy is efficiently converted to light.

This single LED could communicate the words "hello world" by blinking them out as an Internet 0 message (as described in "Communication"). That would be informative for another computer, but not for a person. To write a human-readable hello-world, more than one LED is needed. An image can be decomposed into many small dots, called *picture elements* or *pixels*, with one LED responsible for each pixel. This is how large stadium displays work. Alternatively, it's possible to simplify the jobs of the pixels in a display by asking each to just vary the amount of externally generated light that it reflects or transmits, rather than actively generating light.

The most common way to do this is with a liquid crystal display (LCD). It's not an oxymoron for a crystal to be liquid; the molecules in an LCD are elongated and prefer to line up in the same direction. Scientists call that orientational alignment a kind of crystal, even though the molecules are free to move about as a liquid. Because the molecules can move, an electric field applied across the liquid can rotate the direction in which they point. Light passing through the liquid will try to follow the direction of the molecules, allowing it to be transmitted or blocked based on their orientation. This is how a liquid crystal layer sandwiched between an array of pixel electrodes and light filters becomes a reflective or transmissive display.

Here is the hello-world circuit connected to an LCD, so that now instead of sending text to a computer the circuit can display text it receives from one:

This circuit requires a processor with a few more pins to connect to all of the control inputs on an LCD, but the chip still costs about a dollar and the hello-world circuit is otherwise the same (although, of course, the microcode is different for this application).

The cost of LCDs rises steeply with the number of pixels because it becomes harder and harder to make the displays without some of the pixels being visibly faulty due to manufacturing defects. Another problem with LCDs is that the liquid crystal molecules lose their alignment if the power is switched off, so, like LEDs, power is required to display an image. A number of efforts have tried to get around both limitations by using printing processes to produce displays. One of the most promising is based on microencapsulation. This is the trick that puts ink into carbonless copy paper by enclosing the ink in tiny spheres that break when a pen compresses them. To make a display, the ink in the microcapsules can be replaced with even smaller black and white particles that move in response to an externally applied electric field and then remain where they are when the power is switched off. This is called e-ink; here's what an e-ink hello-world look like:

Note that the display is retaining the image even though it's no longer connected to the controller that powered it. Zooming in, here are what the microparticles are doing in a small part of the image:

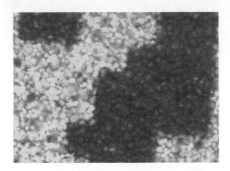

This image shows two kinds of resolution in the display: the small microcapsules containing the ink, and the larger jagged pattern of the pixels from the electrodes that apply the field to move the black and white particles up and down.

One of the cheapest and most versatile ways for a circuit to show an image is to use a ubiquitous display found around the world, from home theaters to mud huts: a TV set. Bought new, TVs are available for a few tens of dollars, and used ones for even less. Until recently, specialized hardware was required to generate the high-speed video signals used by a TV, so they were limited to displaying canned content. But the same kind of microcontroller that blinks an LED to communicate with a person can now blink an electrical signal fast enough to talk to a TV set:

A TV image is made up of hundreds of horizontal lines. The only change to the hello-world circuit is adding a few resistors to produce the voltages that correspond to black, white, and the signal to start a new line. This is what the output from the circuit looks like for one of the video lines, a horizontal cut through the middle of the screen:

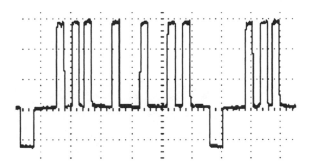

The downward pulse tells the TV to start a new line, the baseline is the black background, and the upward pulses are the white parts of a line through the letters *HELLO*. These pulses are a few millionths of a second long. This format is the NTSC video standard used in the United States, named after the National Television Standards Committee (the standard has also been called "Never the Same Color" because of its idiosyncrasies).

Video standards are a technological mirror of global geopolitics because of the political and economic sensitivity of mass media. As television spread in the 1950s, opposition to a global American standard led to the development of regional alternatives. So, France, the Middle East, and former Communist bloc countries now use SECAM, the Système Électronique pour Couleur avec Mémoire, but more accurately called "System Essentially Contrary to the American Method." And Europe uses PAL, Phase Alternating Line, which has been translated as "Perfection at Last" or "Pay a Lot." The fights over these formats were billion-dollar battles for industrial market share for incompatible products, but as viewed from a one-dollar high-speed microcontroller they're now irrelevant. The same circuit can generate any of these formats by simple changes to the microcode.

For a device to be heard as well as seen, it must be able to make a sound. Compared to generating video signals, which vary millions of times a second, it's as easy as pie (pi?) to produce audio waveforms, which vary thousands of times a second. Here's the hello-world circuit once again, but now connected to a speaker so that it can play out the recording of "hello world" that was made in "Instrumentation":

The samples are streaming from the computer to the microcontroller, and then coming out from a little speaker that's plugged into it.

The speaker is an analog device, responding to continuous voltages. To play digital samples, the processor needs to use a digital-to-analog converter, or D/A, the opposite of an A/D. That conversion is considerably simplified by the tremendous difference in speed between the fast chip and the much slower speaker. The processor can quickly turn a single bit on and off, varying the fraction of time that the bit is on or off. That's called pulse width modulation (PWM). The speaker can't move quickly enough to respond to individual impulses; it filters them and follows the average rate. This is what the digital pulses look like, along with the filtered average of the speaker's continuous response:

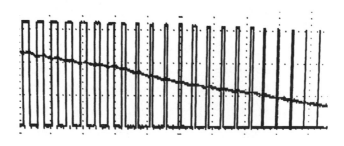

This example uses one of the processor's pins to produce the pulses for the speaker. If more power is needed to drive a bigger speaker, then a simple digital transistor switch can be added to apply the PWM pulses rather than requiring the expensive and power-hungry amplifiers needed for analog signals. The transistor can supply more current than a processor pin because it connects directly to the power supply.

The speaker vibrates because current flowing in a coil on the speaker cone pushes against a magnet on the frame of the speaker. Motors also work by using currents in coils to push on magnets, and there's a particular kind of motor, called a *stepper motor*, that's designed to be controlled by digital pulses rather than continuous currents. These are particularly convenient for connecting to computers.

A stepper motor contains a ring of coils, and within those coils is a ring of magnets with a slightly different angular spacing. When one set of coils is energized, the magnets line up with it. Then when the next set of coils is energized, the motor shaft rotates as the magnets line up with that. Hence the pattern and timing of the pulses to the coils can control the direction and rate of rotation of the shaft without requiring any gears. Here's the hello-world circuit, this time with the transistor switches to provide enough power for the motor's coils:

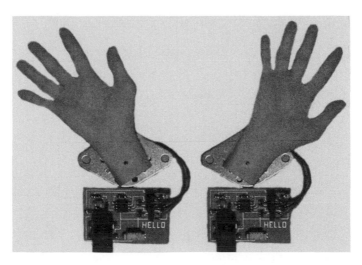

A simple program generates the pulse sequences to set the speed and rotation direction of the motor. With a scanned and printed little hand attached to the motor, this fittingly final hello-world example can wave hello, or good-bye, to the world.

The
Future

The analogy at the heart of this book, between the transition from mainframes to personal computers and that from machine tools to personal fabricators, is not quite correct. The computational counterpart to a machine tool was something even earlier than mainframes: an analog computer. Analog computers solved problems involving continuous quantities, such as finding the angle to aim a cannon in order to hit an artillery target. The variables in a problem, like the distance to the artillery target, were represented by continuous physical properties such as the angle of a gear or the electrical current flowing through a vacuum tube. As the analog computer solved the equations connecting these variables, any imperfections in the shape of the gear or electrical noise in the vacuum tube would lead to errors in the answers, and those errors would accumulate over time so that the longer an analog computation ran, the less accurate the final results would be. The same is true of a milling machine. It continuously positions a cutting tool, and any defects in the shape of the tool or the measurement of its position will cause a permanent error in the shape of the parts produced.

The last serious analog computer was MIT's first computer, Vannevar Bush's Differential Analyzer, built in 1930. Vannevar Bush, then MIT's dean of engineering, would go on to architect the United States' national research and development organization during and after World War II. His student, Claude Shannon, would go on to nearly single-handedly architect the digital revolution.

In the Differential Analyzer, motors powered shafts that turned wheels that corresponded to variables in the continuous (differential) equations defining a problem such as aiming an antiaircraft gun or operating a power grid. By selecting where the wheels rolled against other wheels, and how the shafts connected among them, the relative rates of the rotations would add up to a solution to the problem. In its day this machine was seen as a breakthrough; MIT's head of electrical engineering proclaimed that the Differential Analyzer would "mark the beginning of a new era in mechanized calculus."

We ended up with the Information Age rather than the Mechanized Calculus Age because of some rather severe limitations that Shannon encountered with the Differential Analyzer. It weighed in at a hundred tons. Programming it was an oily process, requiring days to set the ratios of the wheels and the connections of the motorized shafts. Fixing it was an ongoing obligation for Shannon. And even when everything did work, every step in the operation of the Differential Analyzer added an error of about a tenth of a percent.

As Shannon struggled with this unwieldy contraption, it occurred to him that the Boolean logic he had learned as an undergraduate offered an alternative to representing information with wheels. In 1854 George Boole published "An investigation into the Laws of Thought, on Which are founded the Mathematical Theories of Logic and Probabilities," showing that true/false logical propositions could be written and analyzed using algebraic equations. The decision whether or not to carry an umbrella could be written as $R \lor F = U$, where the variable R is true if it's raining, F is true if rain is forecast, U is true if you should carry the umbrella, and $\lor$ is the "OR" operator that is true if either side is true. A more complex Boolean equation could correspond to the logic for a medical diagnosis based on the observed symptoms. Shannon realized that Boole's logical symbols could have a physical counterpart, using the presence or absence of a voltage in a circuit to correspond to a true or false statement, allowing a circuit to solve the discrete equations of logic rather than an analog computer's continuous equations. In 1937 the twenty-two-year-old Shannon wrote what was perhaps the most consequential master's thesis ever, "A Symbolic Analysis of Relay and Switching Circuits," laying the foundation for all of the digital technology to follow.

By 1941 Shannon worked at Bell Labs, investigating the efficiency and reliability of long-distance telephone lines. In 1948 he published in the Bell System Technical Journal "A Mathematical Theory of Communication," uncovering the world-changing implications of communicating digitally. In 1948, the largest communication cable could carry 1,800 simultaneous voice conversations, with ever-increasing demand rapidly outstripping that capacity. Engineers sought better and better ways to use a wire to carry a phone call, never stopping to distinguish between the wire as a medium and the phone call as a message. It was obvious to all that telephone wires were for telephone calls, and the job of a telephone engineer was to connect telephones.

The goal of a telephone was to convert sound waves to electrical waves, send those down a wire, and faithfully convert them back to sound at the other end. Inevitably there would be errors in the reproduction of the sound because electrical noise and interference in the wires would degrade the signal, and the size of those errors would increase with the amount of noise in the phone system:

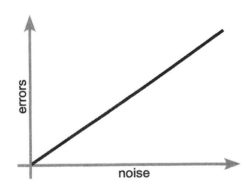

The engineers seeking to improve the phone system would develop ever-more-clever ways to send their signals, so that for a given amount of noise the size of the errors would be smaller:

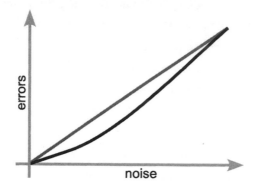

Shannon was able to look beyond those incremental advances and identify the ultimate limits on cleverness. His conclusion had some bad news, and some very, very good news, for the future of the phone system.

He made a profound leap to assume that the signals were being sent in the true/false, on/off, 1/0 digits he introduced into electronics in his master's thesis. His colleague John Tukey called these "bits," as a contraction of "binary digits." This representation not only gave Shannon a precise way to measure the size of the noise and the errors, it anticipated the intermixing of audio and video and text in the Internet. Remember that at the time, digital networks were still far in the future; phone wires were used much like speaker wires to carry audio signals between telephones.

Shannon found that for a digital message the dependence of errors on noise looks *very* different:

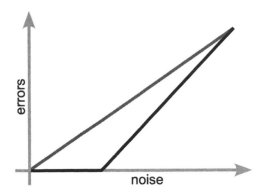

There's a kink in this curve, called a *threshold*. Below a certain amount of noise (which depends on the details of the system), the error rate is effectively zero independent of the magnitude of the noise, and above the threshold the error rate increases again. This threshold is unlike anything that had come before in engineering. Below the threshold, Shannon showed that the imperfect components of a phone system working in the real world could do a perfect job, transmitting information without making mistakes. The appearance of the threshold is the reason why we now communicate digitally,

For given amounts of noise in the system and energy in the signal, the error threshold corresponds to a communication capacity. If the bits are sent below a certain rate (once again, depending on the details of the system), they can arrive without errors, and above that they're guaranteed to have errors. The bad news is the existence of the threshold: there is an end point to the engineering advances in the phone system. Just as the speed of light sets a limit on how fast a signal can travel in our universe, Shannon's result placed a limit on how much information can be reliably sent. But the threshold is also the good news, because instead of engineering improvements leading to smaller and smaller errors, it showed the qualitatively different possibility of making effectively no errors at all.

Error-free communications was realized in the digital communications systems that followed, connecting the globe with signals that could be sent anywhere without degradation by converting sounds to and from digital data rather than retaining the analog waves. And the capacity limit has not proved to be much of a limitation at all; it's been pushed up from thousands to millions to billions and now trillions of bits per second in an optical fiber.

A single bit has to have a digital value of 0 or 1. If, because of noise, it arrives instead with an analog value of 0.1 or 0.9, the error can be identified and restored. If the noise magnitude is great enough to sometimes flip a bit between a 0 and a 1, then the mistake can't be corrected by the bit alone, but a bit-flip error can be corrected if the value of the bit is sent three times and a majority vote is taken because then a mistake will be made only if at least two out of three of the bits get flipped by the noise. More complex schemes can correct errors that occur even more often than that. This is the essence of digital error correction, intentionally adding extra information before sending a signal, and then using that redundancy to identify and correct mistakes.

Shannon's results formed the foundation for what came to be called *information theory*. Until then it wasn't apparent that a theory of information was needed, or could even exist. Information theory grew up in a communications context, but these ideas went on to impact all of the industries that were to eventually use digital technology. Most important they enabled the construction of reliable computers from unreliable components.

In 1945 Shannon's advisor, Vannevar Bush, wrote a prescient essay titled "As We May Think." In it he argued for the application of relays, vacuum tubes, and microfilm to the development of machines that could reason and remember. Although he didn't do so well on the details, his suggestion that computing machines might eventually manipulate ideas with the same dexterity that mechanical machines could manipulate materials was spot-on.

The key to the creation of information-processing machines proved to be the application of Shannon's ideas about communication to computation. His proof of the possibility of perfect digital communications was a singular achievement; the ideas that enabled the subsequent development of perfect digital computers drew on contributions from a larger cast of characters. The recognition of the possibility and implications of error correction in computation emerged in the 1950s from the work of Shannon as well as Warren McCullough, John von Neumann, Samuel Winograd, Jack Cowan, and colleagues.

A computation can be viewed as a kind of communication, sending a message through a computer. In the same way that error correction works on communications between computers by adding extra information and then later removing it in order to catch mistakes and restore the original message, it's possible to error-correct within a computer by performing a redundant computation. Von Neumann analyzed logical circuits that split signals into redundant paths and then combined their results:

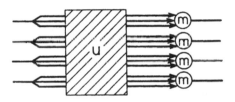

Here, the circles labeled "m" are what he called *majority organs*, which vote on the redundant signals passing through the computing element labeled "u." Von Neumann was able to show, like Shannon, that these kinds of circuits have a threshold property. As long as the error rate in a circuit is smaller than a cutoff that depends on the details of the system, the circuit can produce an exact answer even though all of the components are imperfect. In the same way that Shannon's result led to the development of a theory of information, von Neumann's result formed the foundation for a theory of fault-tolerant computation.

The existence of a threshold in the error rate for computation is what makes computers possible. You might worry about the software in your computer crashing, but you don't have to be concerned about the numbers in a spreadsheet decaying over time. The threshold in information processing is less well known than Shannon's result for communications because its implementation is hidden in hardware, but it's at work in the redundant computers that are used for reliability in aircraft or ATMs. Today's integrated circuits use only the simplest error-correction strategy of ensuring that bits are a 0 or a 1, but as ICs approach fundamental physical limits, fault-tolerant approaches are returning to allow cheaper and more powerful chips to be built by tolerating and correcting errors rather than going to great lengths to avoid them.

Von Neumann's proof of fault tolerance in computing showed that imperfect logical elements could perform perfect logical operations, providing a pathway from the practical limitations of real-world components to the abstract ideal of Vannevar Bush's thinking machines. The question that consumed the rest of von Neumann's life was the study of the essential attribute of life beyond robustness: reproduction. At the end of his life in 1957 he was working on a monograph on the "Theory of Self-Reproducing Automata."

Von Neumann studied automata, discrete mathematical models of machines, because, like Babbage's earlier plan for a mechanical Analytical Engine, his conceptions outstripped the manufacturing technology of his day. He conceived of a machine that could fabricate as well as compute. In its operation an arm would extend to build a duplicate of the machine, then the program to operate the machine would be inserted into the duplicate, whereupon the copy would set about reproducing itself. In the following diagram the Memory Control stores a description of how to build the automaton, the Constructing Unit controls the arm, and the plans for a new Memory Control and Constructing Unit pass through the automata cells of the arm:

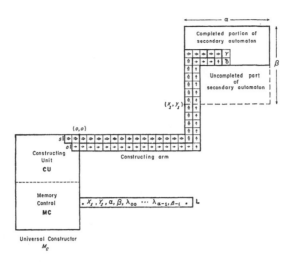

Stripped to its essentials, that's a pretty good summary of our existence. Our cells contain plans for the construction of ourselves, and most of our activities are devoted in some way to the execution of those plans.

Von Neumann studied such a machine as an intellectual exercise, but it's now the subject of serious experimental research. At MIT, in one of the first of a new generation of student theses, Saul Griffith (see "Birds and Bikes") sought to make real mechanical assemblers that could build with materials containing coded instructions. In one version, he developed mechanical tiles that would attract or repel one another based on the arrangement of magnets on their faces. As they are extruded in a linear chain, the chain will fold into any desired shape based on the sequence of tiles, here forming the letters *MIT*:

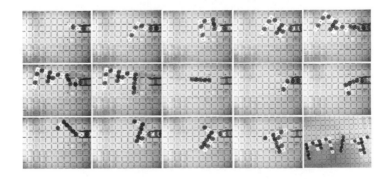

Compare the making of this "MIT" to the processes that are used in what are considered to be the high points of our technological advancement as a species, a ten-billion-dollar chip fab or a jumbo-jet factory. In the chip fab, a great deal of computing goes into designing the chips, and the chips that get built go into computers. But the fabrication process itself is relatively unchanged from what artisans have done for millennia: Layers of materials are deposited, patterned, and baked. The materials themselves are inert. Ditto the NC milling machine that makes structural elements for an airplane; a great deal of computing is used to design the plane and control the tools, and the plane will be filled with embedded computers, but the actual fabrication process happens by a rotating piece of metal whacking away at a stationary piece of metal. The intelligence is external to the tools.

Saul, on the other hand, made his "MIT" in a very different way. He chose the sequence of tiles, and then simply pushed them out. All of the information about what to make was encoded into the materials, so that the description and the fabrication were coincident. This medium is quite literally its message, internally carrying instructions on its own assembly.

Such programmable materials are remote from modern manufacturing practice, but they are all around us. In fact, they're in us. Inside cells there's a remarkable molecular machine called the ribosome:

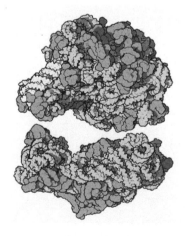

This pair of molecules is a made up of more than a hundred thousand atoms. They're about fifty nanometers across (a nanometer is a billionth of a meter), a thousand times smaller than the width of a human hair. This is where your body builds the proteins that make up your body.

Instructions for the assembly of a protein arrive in the ribosome via a long molecule called *messenger RNA* (mRNA). Messenger RNA is made up of strings of four kinds of smaller molecules, called *ribonucleotides*. These are adenine (A), guanine (G), cytosine (C), and uracil (U). They are grouped in triplets called *codons*, which code for the twenty possible amino acid molecules that proteins are made of. In solution around the ribosome are transfer RNA molecules (tRNAs), carrying each of the types of amino acids.

When an mRNA enters into the ribosome, a tRNA with the matching codon sequence will stick to it. A's stick to U's, and C's to G's:

In this case, the first codon is for methionine (Met). When a tRNA sticks, the ribosome shifts over one position. Then when the newly opened site gets filled, in this case with proline (Pro), the amino acids the tRNAs are carrying get connected by a chemical bond. After that happens the empty tRNA exits the ribosome to go find a new amino acid to carry, the ribosome shifts over one position, and a new amino acid arrives (here, glycine [Gly]). The glycine will get attached to the proline to extend the forming protein (like Saul's growing chain of tiles), the ribosome will shift one more position along the mRNA, and this procedure will continue until all of the amino acids are attached to assemble the molecule. Once a molecule is strung together it leaves the ribosome, whereupon the molecule folds into its final shape, which will determine its function, just as Saul's letters folded. This is how your proteins are made, including the ribosomes themselves—a ribosome can make itself.

Think as Shannon or von Neumann might have about how the ribosome builds a protein, rather than as a molecular biologist might. First, the genetic code is redundant. There are 4 x 4 x 4 = 64 possible nucleotide triplets, but they code for only twenty amino acids. Some of the remaining available codons are used for control functions such as marking when to stop making a protein, but the rest are a redundant code that reduces errors. Then, there's a proofreading mechanism to ensure that tRNAs attach to the right amino acids, and as the amino acids are added the ribosome can catch some errors in its own operation. Finally, proteins that don't fold correctly can be degraded by the cell. All of this means that the chance of the ribosome making a mistake in assembling a protein is just one out of about every ten thousand steps. Another molecule, DNA polymerase, works like the ribosome but has the even more sensitive job of copying DNA when the cell divides. It adds an extra error-correction operation to check and repair its work, giving an error rate of less than one step in a billion. This is an extraordinary number! These molecules are randomly jostling around in the messy environment inside a cell, yet their reliability would be the envy of any manufacturer. This is accomplished by using logic to build, by running a molecular program based on executing coded genetic instructions. By even the narrowest definition, the ribosome and DNA polymerase compute.

There's a pattern here. Shannon showed that digital coding can allow an imperfect communications system to send a message perfectly. Von Neumann and colleagues showed that digital coding can allow imperfect circuits to calculate perfect answers. And the ribosome demonstrates that digital coding allows imperfect molecules to build perfect proteins. This is how the living things around you, including you, form from atoms on up. It's necessary to precisely place 10^{25} or so atoms to make a person, an ongoing miracle that is renewed in everyone every day. The role of error correction in fabrication is as close as anything I know to the secret of life.

The very close analogy between the role of digitization in reliable communications and computation and its role in fabrication in the ribosome suggests a corresponding analogy between the prior revolutions in digital communications and computation and a revolution to come in digital fabrication. Following that analogy, this will be a revolution in manufacturing technology, based like the ribosome on materials that contain codes for their organization, and tools that build by computing. The role of computation isn't an afterthought in the ribosome; it's intrinsic to its operation. Unlike the option of external computer control of a milling machine, the computer can't be separated from the ribosome—the ribosome *is* the computer. Today's most advanced factories use digital computers to manipulate analog materials; in digital fabrication the intelligence is internal to the assembly process, offering exactly the same benefits for fabrication as the digitization of communication and computation. Imperfect machines will be able to make enormously complex perfect parts.

To even call digital fabrication a manufacturing technology risks trivializing it, just as the original dry descriptions of thresholds in circuits for communication and computation hid their eventual implications in the creation of Web sites and portable music players and video games. Trying to predict the uses of programmable fabrication is likely to be as unsuccessful as it would have been to ask Shannon or von Neumann to forecast the future uses of networks or computers. But the historical analogy between the digitization of computation and fabrication does suggest the nature of the eventual answer: personalization. I might want a portable personal space for screaming, you might want an analytical agricultural implement. The real impact of digital communications and computation came in giving ordinary people control over the information in their lives; digital fabrication will likewise give individuals control over their physical world by allowing them to personally program its construction.

The earlier information technologies followed from the communication and computation threshold results because they made possible engineering on a scale of unprecedented complexity. Networks and computers grew to contain more components than had ever been built into mechanisms before, with the last part operating as reliably as the first one. Likewise, digital fabrication will enable perfect macroscopic objects to be made out of imperfect microscopic components, placing not just 3 or 10^3 but ultimately 10^{23} parts. ($6.023 . . . . \times 10^{23}$ is Avogadro's number, used by scientists as a unit to count the microscopic constituents of macroscopic things; it's equal to the number of molecules in a cubic centimeter of an ideal gas, and was named after the Italian chemist Amedeo Avogadro who in 1811 hypothesized that equal volumes of gases at the same temperature and pressure contain equal numbers of molecules.)

A new kind of engineering design practice is going to be needed for this limit of Avogadro-scale complexity: A conventional CAD file specifying where to put 10^{23} parts would be about as large as the thing to be made. Even worse, based on the current difficulties engineers face with unexpected behavior in engineered systems with millions or billions of components, it's unlikely that something with that many parts would do what its designers expect. One of the intellectual frontiers associated with the development of digital fabrication is the development of an associated design theory that can specify what an enormously complex machine should do without specifying in detail how it does it. The challenge is to do this new kind of large-scale engineering design with the same kind of rigor that allows engineers today to specify the performance of an airplane or computer chip. Attempts to manage engineering complexity by emulating biology have not been all that successful, because they've generally been limited to anecdotal examples—a biologically inspired circuit or structure may work on a toy test problem, but there's no guarantee that it will continue to work on challenging real-world tasks. Instead, there are intriguing hints that it's possible to forward- rather than reverse-engineer biology, rigorously deriving the design of biological-like systems from a precise problem statement and clear understanding of the purpose of an engineering design. Just as Shannon showed how Boole's logical equations could have a physical representation in a circuit, these emerging insights into Avogadro-scale engineering are showing how to create physical representations of equations called *mathematical programs,* which express goals and constraints.

The possibilities of building with logic and programming with math are contributing to the emergence of a theory of digital fabrication, as a mate to the earlier theories of digital communications and computation. In the laboratory these ideas are being embodied in research projects developing programmable fabricators, with approaches based on both building up molecular assemblers and shrinking down mechanical analogs like Saul's tiles. These will work something like the ribosome, but because they're not restricted by biological evolution they can work beyond the ribosome's limits on the available types, sizes, and specifications of materials and structures that can be created, able to use nonbiological materials like steel and semiconductors.

Programmable digital fabricators will eventually make it possible to drop the "almost" from my class "How To Make (almost) Anything," offering in the physical world exactly the same kind of universality provided by a general-purpose computer. Actually, "How To Make Anything" won't quite be right either; the correct title will have to become "How To Make (and unmake) Anything," because the inverse of digital fabrication is digital recycling. Just as one can sort a child's LEGO bricks back into bins, or the bacteria in a compost pile can disassemble organic waste back into its molecular constituents, digital fabrication processes can be reversible, with the tool that builds something able to unbuild it. Trash is an analog concept, based on an implicit assumption that the means to make something are distinct from the thing itself. But the construction of digital materials can contain the information needed for their deconstruction. Digital computers already "garbage collect" to recover memory filled by unneeded bits, and digital (dis)assemblers will be able to do the same with unneeded objects to recover and recycle their raw materials.

Interestingly, digital fabrication was almost, but not quite, invented in the 1950s. John von Neumann's mathematical model of a self-reproducing machine is a discrete automaton, using then emerging ideas for digital computation. At the same time, the same people who were thinking about self-reproducing machines were also thinking about fault-tolerant computers. And von Neumann wrote presciently on "The Role of High and of Extremely High Complication," about the difference in complexity attained in engineered and biological systems. But there's no record of these ideas ever having been connected back then, in a proposal for making perfect things out of imperfect parts by building through computation.

Actually, the invention is a few billion years old, dating back to when the biological machinery for protein translation first evolved. Today, the theory of fabrication that is emerging is, like any good scientific theory, descriptive as well as prescriptive. It helps explain the central role of coding and error correction in how biology builds, and helps guide the development of nonbiological fabricators capable of operating beyond biology's limits.

Not that there will be much difference left between biology and engineering at that point. Beyond the fascinating (and contentious) history of the philosophical and religious debate over the definition of life, viewed purely phenomenologically a computing machine that can duplicate, program, and recycle itself has the essential attributes of a living system. Von Neumann studied self-reproducing automata in order to understand life; progress in that understanding is now leading up to the creation of something very much like life. Once a machine becomes capable of reproducing itself, there's not much difference left between personal fabrication and the fabrication of a person.

Joy

In 2000, Sun Microsystems' chief scientist Bill Joy published an intelligent and influential essay in *Wired Magazine* titled "Why the Future Doesn't Need Us." In it he argued that the development of self-reproducing molecular assemblers of the kind I described in "The Future" would be so dangerous to our survival as a species that we should consider voluntarily abandoning the effort. In considering the future of personal fabrication, it's indeed worth thinking about where it's headed, and why, as well as how it will get there. The state of the technology today does provide intriguing insight into whether personal fabrication will improve or destroy the world of tomorrow.

Technologists do have a terrible track record for reflecting on the wisdom of where their work might lead. Bill Joy's sensible point was that robotics, genetic engineering, and biotechnology are so fundamentally and dangerously unlike any technology that has come before that we can no longer afford to blindly develop them and hope that it will all come out OK. There are two reasons to see these technologies as different from all the rest. First, the only barrier to entry in each of these areas is access to knowledge, not scarcity of resources. It requires an enormous and enormously expensive supply chain to produce plutonium to make a nuclear weapon; neither the machines nor the money are easily hidden. But producing a protein just requires easily available reagents and a laboratory bench to mix them on. Ideas are so powerful precisely because they can't be controlled in the same way that scarce physical resources can.

And second, genes and robots can make more genes and robots. Self-reproduction, as I described it in the last chapter, implies exponential growth. One bacterium can divide and make two bacteria, which divide and make four, then eight, then sixteen, on up to a plague. If a new technology is truly capable of self-reproduction, then it could cause a technological plague, growing without bound until it consumes all available resources on earth.

Eric Drexler, a pioneer in nanotechnology, termed this the "gray goo" scenario in his 1986 book *Engines of Creation: The Coming Era of Nanotechnology* refers to technologies working on the scale of nanometers, 10^{-9} meters, the size of atoms. Eric's colorful (colorless?) description of unchecked self-reproducing nanorobots as a gray goo follows from the likelihood that they would appear to the naked eye as a featureless mass of growing stuff.

This frightening prospect has led to opposition, sometimes violent, to unchecked nanotechnology research from unlikely bedfellows, from Prince Charles to environmental activists to survivalists. If self-reproducing machines could take over the planet, and if the only restraint on this happening is control over the spread of knowledge about how to do it, then we're doomed because ideas can't be contained. Whether out of malicious intent, benign neglect, or misplaced optimism, a single technological Typhoid Mary could destroy everyone and everything.

Rather, we will be doomed if that does happen, but there are a number of reasons for thinking that it might not. It is true that self-reproducing nanorobots are fundamentally unlike earlier technological threats to our survival because of the risk of exponential growth, and because of the difficulty of containing a technology based on ideas rather than rare raw materials. Then again, each of the other end-of-life-on-earth risks have been fundamentally different from their predecessors. Firearms brought an immoral edge to warfare, making it possible to kill at a distance, and rendering existing armor obsolete. Then came nuclear weapons, which could erase whole cities rather than just kill individuals. More recently, the viability of the networked global economy has been challenged by computer virus writers, with lone hackers able to cause billions of dollars' worth of damage by remotely shutting down networks and computers.

We've survived so far because of coupled advances in evaluating these threats and developing countermeasures. Firearms are as deadly as ever, but police and soldiers can now wear Kevlar and ceramic body armor that provides protection, which would have been inconceivable to a blacksmith. For nuclear weapons, the attempt to build a Star Wars–type shield has remained a very expensive kind of science fiction, but the combination of a nuclear deterrent and nonproliferation regimes for monitoring and controlling nuclear materials has provided a historically unprecedented restraint on the use of such a fearsome military weapon. At the dawn of their creation, nuclear weapons represented an almost unimaginably large force; at the Trinity test of the first nuclear weapon, General Groves was annoyed by Enrico Fermi's offer to take wagers on whether the test would ignite the earth's atmosphere, or perhaps just New Mexico's. Hans Bethe was able to show that atmospheric ignition can't happen, and predicting the yield of a nuclear weapon has now become a rather exact science. It would likewise be a mistake to assume the use of twentieth-century technologies in analyzing and addressing a twenty-first-century threat.

Technologies coevolve with their design tools and countermeasures. Computer virus writers found vulnerabilities in the Internet and in operating systems; a combination of software engineering (firewalls, software patches, e-mail filters) and social engineering (financial liability, legal prosecution) has not eliminated computer viruses but has turned them into a manageable nuisance. The same is true of the existing example of a self-reproducing machine, a real virus like AIDS.

What's surprising about the AIDS epidemic is not the scale of human suffering it's already caused, horrific as that's been; it's that the epidemic hasn't been even worse. The HIV virus that causes AIDS is a fiendishly clever mechanism. HIV is a nearly pure form of information. At a size of about a hundred nanometers it's too tiny to move far on its own or to carry much in the way of energy; it obtains its resources by hijacking a host cell. HIV does this with three enzymes. The first of these, reverse transcriptase, converts HIV's RNA into DNA. The second, integrase, splices the DNA into the host cell's own DNA. When the cell translates this DNA into a protein, the third enzyme, protease, cleaves the protein to assemble a new virus. This process is so effective precisely because of the reliability of the copying done by DNA polymerase. DNA replication is so carefully controlled that the cell trusts the information in DNA and doesn't bother to check for or protect against "unauthorized" tampering. Reverse transcription runs backwards from the usual direction of transcription from DNA to RNA, and because the virus doesn't error-correct its reverse transcription this introduces a mechanism for rapid mutation that helps the virus evade our natural immune response. But it also introduces a vulnerability into the life cycle of the virus that's been a successful target for our technological immune response, through AIDS drugs that disrupt reverse transcription.

HIV embodies our most advanced insights from biology, chemistry, physics, and computer science. It's hard to believe that we could accidentally or intentionally come up with a self-reproducing machine even more effective than HIV, and even if we did it would have to compete for natural resources with all of the other viruses and bacteria on turf they've spent a few billions years evolving to optimize their designs around. Molecular machines will confront the same challenges in obtaining energy, eliminating waste, and defending against threats that limit the growth of their biological predecessors.

Eric Drexler came to regret coining "gray goo" as a distraction from the far more likely development of nanotechnology based on programmed molecular manufacturing. Nanotechnology itself is a curious sort of technology. It's the subject of billion-dollar research strategies and breathless press coverage, but very few of the serious scientists studying small things call their work nanotechnology. Molecular biologists are developing minimal cells, either from scratch or by simplifying existing cells, to create controllable bioreactors for producing drugs and chemicals. Chemists are studying new materials that self-assemble from molecular building blocks. Device physicists are developing more powerful information technologies based on manipulating individual atoms, electrons, and photons. The arrival of atomic-scale digital fabrication will not be a discontinuous event; it's already here in these rudimentary forms, which will continue to emerge and converge at the intersection of these disciplines.

Planetary annihilation from the unchecked growth of self-reproducing molecular machines may be an unlikely doomsday scenario, but if molecular fabricators that can truly make anything are indeed coming, then a far more practical fear is who might make what. When I first briefed a roomful of army generals on personal fabrication and digital assembly, that question triggered a lively discussion. One contingent was appalled that we were intentionally fostering access to personal fabrication through fab labs, putting these tools in the hands of potential attackers, and they muttered about the need to control this technology. Another contingent argued that it wasn't a controllable technology, and in any case the bad guys would get their hands on it regardless of any attempted limits, and that the impact could be far more beneficial in helping address the root causes for conflict. Guess which side I was on.

Bad guys are already impressively effective at acquiring the best available technology for the destruction of their enemies; fab labs are likely to have a far greater impact on the stability of the planet by helping everyone else acquire the technology they need for their survival. Debating the wisdom of widespread access to personal fabrication tools reflects a divergence in the discussion of the appropriateness of

technology between people whose needs are and are not well met, a kind of intellectual version of not-in-my-backyard.

I first heard an early echo of the chorus of voices in this book in the response to my earlier book *When Things Start to Think*, which was about moving computing out of traditional computers and into everyday objects. On one hand, I was happily swamped with out-of-the-blue letters along the lines of "Thank heavens, finally this is information technology that's relevant for me, I need this for. . . ." They would then go on to describe their ideas for applications of embedded computing in their lives. On the extreme other hand, I was startled to read in a review that the book was irrelevant because of the well-known impossi-

bility of computers thinking. Aside from rehashing an unwinnable old philosophical argument about defining intelligence, my learned reviewer had entirely missed the point. Embedding computer chips was not an exercise in epistemology; it was a way to solve problems. As the tools for personal fabrication now make accessible not just the use but also the creation of technology, users rather than pundits can decide which problems need solving.

It's common to bemoan how kids no longer have hardware hobbies, and, as they get older, how so few are interested anymore in technical careers. These may have been temporary casualties of the digital revolution, artifacts of an overemphasis of bits over atoms as young and old sat before their computer screens. Kids' toys emulate grown-up tools; a child's play with an erector set was not too far off from the work of an

engineer. More recently, other than writing software it hasn't been feasi-ble in the home to do anything remotely close to the frontiers of sci-ence and engineering—the required tools became too expensive and specialized. Now, though, that pendulum is swinging back so far that a child playing with personal fabrication tools can have access to capabili-ties beyond those available to an individual engineer working as a cog in the gears of an organization based around central scarce technical resources.

And play the kids do. In this book, I've described a deep idea: the digitization of fabrication. I've explored its implication in personal fab-rication. I've presented the observation that we're already in the equiv-alent of the minicomputer era of personal fabrication. But the most unexpected part of this story has been the response, as the nearly universal pleasure in getting access to tools for technological develop-

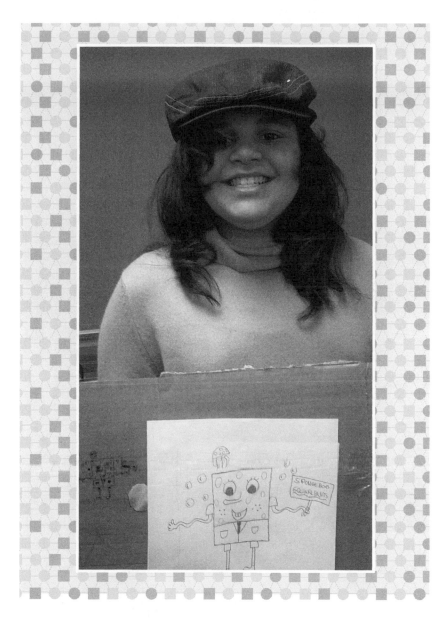

ment appears to cut across ages and incomes, from tribal chiefs to MIT students to children coming in off the street. There's a shared childlike delight in invention that's shown best of all by a child.

The liberal arts were originally meant to liberate, empowering through mastery of the means of expression. The new means of

expression for personal fabrication offer technological empowerment, providing economic, intellectual, and even a kind of spiritual liberation. I believe that the best place to read about the future of personal fabrication is in the faces of a new generation getting access to prototype versions of these capabilities in fab labs. Coming from many different parts of the planet, they end up in the same place: joy.

The
Details

This was not an easy book to write. During its incubation financial bubbles burst, wars were won and lost, and societies and economics transformed. Against this backdrop of a world of change, the story of personal fabrication emerged for me as a personal trajectory through some of the most exciting and demanding places in our world, and in the world of ideas.

Operationally, personal fabrication cuts across existing institutional boundaries. As both a cause and an effect of that, work on it has been associated with institutional changes that proved to be at least as challenging as the research itself. At MIT, the Center for Bits and Atoms was growing from its roots in the Media Lab and into a campuswide program. Beyond campus, the progenitors of fab labs grew out of the Media Lab Asia project in India, which nominally had nothing to do with personal fabrication. Each of those organizations contributed to the story told in *Fab*, in often unplanned ways.

Intellectually, *Fab* is the synthesis and culmination of three earlier books that I wrote, even though it wasn't at all apparent to me then that those were leading up to this now. *When Things Start to Think* looks at how and why computing can move outside of conventional computers and into the world, *The Physics of Information Technology* covers physical mechanisms for manipulating information, and *The Nature of Mathematical Modeling* contains mathematical methods for describing worlds both real and virtual. Taken together, these add up to a prescription for programming atoms as well as bits, and hence point to the possibility of digitizing and personalizing fabrication as well as computation.

Unlike those earlier books, though, *Fab* presents things as well as ideas. The "writing" of *Fab* entailed authoring not just text but also CAD files, circuit diagrams, PCB layouts, and computer programs for processors big and small. Likewise, *Fab*'s "readers" have included a small army of fab class students and fab lab users, collaborating colleagues and friends, and a heroic review of the text itself by CBA's program manager Sherry Lassiter and Shelly Levy-Tzedek.

Fab would not have appeared if not for the patient editing and thoughtful guidance by Jo Ann Miller, Iris Richmond, and their colleagues at Perseus and Basic Books, throughout an ambitious if not always timely writing project.

And *Fab* would not have been written, or been worth writing, if not for the unexpected global growth of the fab labs, which has been one of the most rewarding activities I've ever been involved in. In doing this, I've been fortunate to have had a tremendous group of smart and capable collaborators, including Sherry Lassiter in the office as well as enthusiastically directing the field fab lab activities in our backyard and around the world, CBA's able administrator Susan Murphy-Bottari, our lead NSF program managers Kamal Abdali and Denise Caldwell along with Mita Desai in the original launch, and MIT students including Amon Millner, Aisha Walcott, Chris Lyon, Caroline McEnnis, Reginald Bryant, Han Hoang, Raffi Krikorian, Manu Prakash, Yael Maguire, Ben Vigoda, Jason Taylor, Michael Tang, Jon Santiago, and the unstoppable Amy Sun. Of the fab labs far enough along to be mentioned in the book, the Indian labs emerged from memorable discussions on long trips bouncing along India's back roads with Drs. Sanjay Dhande, now the director of the Indian Institute of Technology in Kanpur, and Debu Goswami, now of the IITK faculty; the introduction to Haakon Karlsen and Norway's Lyngen Alps came via Telenor's Pål Malm, and the fab lab launch was assisted by Telenor's Bjørn Thorstensen and Tore Syversen, and Sami Leinonen from UPM-Kymmene; Bakhtiar Mikhak initiated the lab in Costa Rica with Milton Villegas-Lemus; MIT's Bill Mitchell made the connection to Mel King and Boston's South End Technology Center, and that lab has been helped by an extended family including Ed Baafi, Joel and Gabe Cutcher-Gershenfeld, and Chris and Wyatt Cole; the SETC lab was soon followed by a Ghanaian connection through Robert Baafi, Renee Dankerlin, Kofi Essieh, and Reginald Jackson.

I've relegated to here, the last part of the last part of the book, the hardware details of the models and manufacturers of the equipment that I used for the hello-world examples in *Fab*, since that information may be timely but is far from timeless. While it will almost instantly be obsolete, my hope is that it will remain valuable as a starting point for keeping current with the state of the art. The engineering files used with this hardware to make the hello-worlds are posted online at http://fab.cba.mit.edu/fab, along with an inventory. Taken together, these provide enough information to reproduce the examples in the book, and, even better, to improve on them.

The Past

My interest in the origin of the ideas behind computing machines grew out of time that I spent at Bell Labs, Harvard, IBM, and MIT with some of the early computing pioneers and their collaborators, who we're fortunate to still have around. The AI guru Marvin Minsky, who studied with John von Neumann, has in particular been an invaluable guide. Looking further back, I learned about the connection between makers of scientific and musical instruments from the historian Myles

Jackson, author of *Harmonious Triads: Physicists, Musicians and Instrument Makers in Nineteenth-Century Germany*.

Hardware

My foray into rapid prototyping of software for rapid-prototyping hardware was made possible by the Python programming language, created by Guido van Rossum, and the many graphical, mathematical, and I/O modules that have been written for it. Python gets passed on like a meme, causing its converts (like me) to babble incoherently about the virtues of the language. The still-developing CAM program used in the example, cam.py, is linked off of http://fab.cba.mit.edu/fab.

Subtraction

The tools used in this chapter ranged in cost from a few thousand dollars for the tabletop mill (Roland Modela MDX-20) and sign cutter (Roland CAMM-1 CX-24), to tens of thousands of dollars for the laser cutter (a 100-watt Universal X2-600) and machining center (Haas Super Mini Mill), and approaching a hundred thousand dollars for the waterjet cutter (Omax 2626). The Roland machines and a smaller laser cutter (a 35-watt Epilog Legend 24TT) formed the heart of the first fab labs, embodied in the machines themselves as well as the spirit of the companies behind them.

Addition

It was, as always, a pleasure to work with CBA's able shop technician John DiFrancesco on the examples in this chapter. The 3D printing was done with a Z Corp Z406, the injection molding with a WASP Mini-Jector #55, and the vacuum forming with a Formech 660. An example of a printing process for semiconductor logic as described at the end of the chapter is: *All Inorganic Field Effect Transistors Fabricated by Printing*, B. Ridley, B. Nivi, and J. Jacobson, *Science* 286 (1999), p. 7469.

Building Models

The building-scale printing effort that I refer to is led by Berok Khoshnevis at the University of Southern California, who is developing mobile analogs of 3D printers that can pump concrete.

Description

The bitmap drawing program used in this chapter was GIMP, the 2D drawing program was drawn from OpenOffice, and the 3D programs were Blender and Cobalt. The 3D scanner was a Konica Minolta VIVID 910.

Computation

The schematics were drawn and PCBs laid out with the program Eagle. The microprocessors used in *Fab* are the Atmel AVR family, including the 8-pin ATtiny13 and ATtiny15, and the 20-pin ATtiny26 (in the low-power low-voltage L and V versions, and the higher-speed higher-voltage variants). These squeeze A/D, PWM, timers, comparators, amplifiers, EEPROM, and FLASH memory around a high-speed RISC core in a tiny package that sells for about a dollar, making possible the range of fab lab projects. In-circuit programming was done by the UISP program via a parallel-port cable, using code generated by the AVRA assembler and GCC compiler. The subtractively machined circuit boards were made out of paper-based CEM instead of the more common glass-based FR4, for longer tool life. The 1/64-inch center-cutting end mills came from SGS Tools (available discounted through bulk resellers). The flexible cut circuits used 3M's #1126 copper tape with electrically conducting glue, #214 transfer adhesive, and #1 epoxy film as a substrate. They were cut with tools from Roetguen's "easy weed" blades, once again, available through bulk resellers. Electronics components were bought from Digi-Key's marvelous Web site, and mechanical components from McMaster-Carr.

Instrumentation

The general case of "seeing" with electric fields mentioned at the end is discussed in Josh Smith's 1999 MIT Ph.D. thesis "Electric Field Imaging," implemented in Freescale Semiconductor's field-imaging line of chips, and commercialized by Elesys for automotive occupant sensing.

Communication

Internet 0 is described in *The Internet of Things,* Neil Gershenfeld, Raffi Krikorian, and Danny Cohen, *Scientific American* 291 (October 2004), pp. 76–81.

Interaction

The electronic ink hello-world was done by Ara Knaian at E Ink Inc.

The Future

This chapter, and my thoughts about the future, benefited from lively discussions with CBA colleagues including Saul Griffith, Joe Jacobson, Ike Chuang, Seth Lloyd, and Shuguang Zhang. The Ribosome image appears courtesy of David S. Goodsell of the Scripps Research Institute, from his beautiful "Molecule of the Month" collection for the Protein Data Bank.

Joy

The images in this chapter are examples of a nearly limitless supply of pictures of beaming people that my colleagues and I have accumulated while documenting work in fab labs. The first picture shows a typical scene after the opening of the Ghanaian fab lab at the Takoradi Technical Institute (TTI). One seat would invariably be occupied by many bottoms, as kids of all ages collaborated on mastering and using the tools. In this picture, they're learning to prepare designs for the laser cutter. The second shows Haakon Karlsen's son Jørgen peeking out from behind the first mesh 802.11 (Wi-Fi) antenna that he made with Benn Molund for the electronic shepherd project. The third is Valentina Kofi, an eight-year-old who fell in love at first sight with electronics at the TTI fab lab. She's holding the first circuit board she made, an "Efe." *Efe* means "it's beautiful" in the Twi language; the circuit board was developed in the fab lab to show how to interface a touch sensor and an LED with a microcontroller communicating with a computer. This was taken after she had stayed late into the night to get it working. The fourth picture was taken after an unexpected bit of serendipity. Bitmap image input had just been added to the fab lab CAM program, and independently a scanner had been brought into

Boston's South End Technology Center (SETC) fab lab. We then realized that these could be used together to scan a drawing on a piece of paper and turn it into a toolpath for the laser cutter to output the drawing in another material. We tried this first on Meleny's sketch of SpongeBob, rescaling it and cutting it into a piece of cardboard. Meleny's delight was immediately followed by her asking if she could get a job doing this. That question was one of the inspirations for the high-tech craft sale that I mentioned in ". . . Almost Anything." The final image shows Dominique, one of the first girls at SETC to discover the joy of doing arts-and-crafts projects with the rapid-prototyping tools, here cutting out her name on the sign cutter.

Finally, demand for fab labs as a research project, as a collection of capabilities, as a network of facilities, and even as a technological empowerment movement is growing beyond what can be handled by the initial collection of people and institutional partners that were involved in launching them. I/we welcome your thoughts on, and participation in, shaping their future operational, organizational, and technological form.

Index

Abdali, Kamal, 258

Acoustics, 36

Acrylic, 69, 70, 130

A/D. See Analog-to-digital converters

Additive technologies, 91, 93–119

Adinkra symbols, 80

Aerospace industry, 106, 117

Agriculture, 82, 88–92, 173
    precision agriculture, 164–165

AI. See Artificial intelligence

AIDS, 245–246

Air conditioning, 87

Alarm clocks, 23–24

Alda, Alan, 62, 63(photo), 64

American Standard Code for
    Information Interchange.
    See ASCII code

Amino acids, 10, 239, 240

Amplifiers, 225, 261

Amponsah, Kyei.
    See Nana Kyei

Analog-to-digital converters (A/D),
    173–176, 178, 261

Analytical Engine, 36

"An investigation into the Laws of
    Thought…" (Boole), 230

Antennas, 12, 68, 182

Ants, 149–150

Apple II, 9, 137

Applications, 9, 49, 51, 148, 175,
    195, 201, 204
    killer apps, 9, 144

APT. See Automatically Programmed
    Tools

Architecture, 6, 103, 104–110, 114,
    119, 190–195, 209
    and engineering design and
        production tools, 109–110
    and functional infrastructure, 192
    and information technology, 190

Art/artists, 6, 8, 31–33, 34, 36, 55,
    104, 119, 131
    and use of technological tools, 114

Artificial intelligence (AI), 48, 133,
    134, 136, 191

Artisans/artifacts, 8, 31, 32, 34,
    36, 55

ASCII code, 197–198, 202

Children, 26–27, 80, 133, 134–135,
139, 142, 143, 145, 146(photo),
147, 163, 211–212, 214, 252–254,
263
Buddhist and Muslim girls, 216
David (inventor). 140–142,
141(photo)
Chips, 9, 11, 114, 139, 155, 170,
174, 202, 224, 238. See also
Microcontrollers; Microprocessors;
under Costs
Chlorine gas, 69
Christiansen, Godtfred Kirk and Ole
Kirk, 137
Chuang, Ike, 144, 262
Church, Alonzo, 37
Church-Turing thesis, 38
Circuits/circuit boards, 9, 11, 12, 21,
27, 49, 68, 76, 114, 117, 118, 174,
175, 203, 219, 230, 236, 240, 241,
243, 251, 262
schematic diagrams, 151–152
fabricating, 25, 26, 50, 101
in-circuit programming, 160
and LCD, 220–231
printed circuit boards (PCB), 153
sample-and-hold circuits (S/H),
178–179
thermal noise, 175
through-hole and surface-mount
parts for, 154–155
Cities, 35, 82
C language, 157
Clay, 104, 129
Coal, 35, 90
Color, 99
Communication(s), 181, 187, 190,
195, 197–205, 216, 219, 234, 241
communication capacity, 233, 234
error-free, 233–234
between machines, 197

serial/parallel, 198–199
See also Telecommunications
infrastructure
Communist block countries, 223
Community leaders, 77, 91, 207
Comparators, 175, 176, 261
Compilers/compiled languages,
156–157
Complexity, 194, 195, 242, 244
Computation, 4, 10, 91, 149–160,
234–235, 241, 243
as communication, 235
computationally-enhanced
materials, 195, 238, 241
error correction in, 235. See also
Error/error correction
fault-tolerant computation, 235,
236
personalization of, 8, 14, 41
Computer aided design (CAD), 41,
44(illus.), 105, 106, 114, 117, 120,
122, 124, 131, 217
limitations of, 129
and modeling behavior, 128–129
operating modes of, 44–45
parametric programs, 127—128
3D programs, 125–126
Computer aided manufacturing
(CAM), 40–41, 45, 48, 49, 50, 260
universal CAM tool, 50
Computers, 84, 105, 190, 193, 243
analog computers, 229
assembling from discarded
computers, 79
computer communication interface,
152
computer games, 25
computer languages, 17, 40, 135,
156–159
computer networks, 15
computer viruses, 247